Quest for the Invisibles

by

Nik Hayes

THE BOOK TREE
San Diego, California

© 2016
Nik Hayes
All rights reserved

No part of this publication may be used or transmitted in any way without the expressed written consent of the publisher, except for short excerpts for use in reviews.

ISBN 978-1-58509-149-2

Cover layout and design
Mike Sparrow

Cover photos copyright by
Nik Hayes

Published by
The Book Tree
P O Box 16476
San Diego, CA 92176
www.thebooktree.com

We provide fascinating and educational products to help awaken the public to new ideas and information that would not be available otherwise.
Call 1 (800) 700-8733 for our *FREE BOOK TREE CATALOG*.

Dedicated To

Trevor James Constable

September 17th 1925 – April 2nd 2016

During the final stages of finishing this book and getting it ready for publishing I was given the sad news by Paul Tice of the passing of Trevor James Constable on April 2nd 2016. Being a musician for much of my life I have been influenced by a few people over the years, but none more so than Trevor. After initially reading his classic book *The Cosmic Pulse of Life* back in 2009, I wrote to him saying just how much of an affect it had on me, and how I had now decided to start my own quest to photograph invisible UFOs. He wrote back to me with kind and encouraging words, and it was this reply that really urged me to go forward and begin my work. Trevor James Constable was way ahead of his time, and in my opinion the world owes him a great debt of gratitude for his pioneering UFO photography. I present this book in honour of his life and his work.

Acknowledgements

The author would like to thank the following people: All my family and friends for their constant support; Paul Tice at The Book Tree for all his support and encouragement, and for helping to make the dream a reality; Ellie James for all her help and encouragement; Dan Llewellyn at maxmax.com for the camcorder conversions and filters, and for providing a great service; Gregory Harold for his support, as well as for the video and books; Karon Annells, Alan Evans, Donna Evans, Bill Brace, and everybody else who has helped and supported me along the way.

Contents

Acknowledgements..4

Introduction 7

Chapter 1 A New Reality...9

Chapter 2 The Search for Answers.....................................19

Chapter 3 Applying the New Knowledge.........................29

Chapter 4 Skyfish..42

Chapter 5 Spinning Jennys and Plasmatic Gliders.........52

Chapter 6 A Glimpse into the Ultraviolet........................62

Chapter 7 Back to Normality..74

Chapter 8 Techniques for filming Invisible UFOs.........85

Chapter 9 Conclusions...98

Appendix 1: Infrared Photographs of Invisible UFOs..................116

Appendix 2: Ultraviolet Camcorder Stills of Invisible UFOs........129

Appendix 3: Ultraviolet Camcorder Stills of Skyfish....................141

Bibliography...151

Introduction

From a technological standpoint there is no better time in history for looking into this invisible world, than right now.

—Trevor James Constable

On Christmas day evening 2008 whilst out walking one of my dogs, my attention was drawn to a craft-like object which was travelling at very low altitude, and moving at little more than walking pace. It was mostly hidden by the thick freezing fog which covered much of Oxfordshire at the time, but as it grew nearer I could see a faint reddish glow which gradually became more and more intense. As it proceeded to move through a much thinner band of fog I managed to get a good view of it as it passed directly overhead.

This low altitude encounter convinced me that there must be some truth to the UFO enigma, and after a lot of effort I finally managed to capture an image of it using the movie mode function on my digital camera. I then became more and more transfixed with the idea recording more examples, and ended up spending most of the next year researching the subject, as well as watching as many UFO documentaries as I could find on the Internet. This is where I first discovered the pioneering work of Trevor James Constable, and immediately began searching for any books he may have written.

I came across *The Cosmic Pulse of Life*, which had recently been reissued, and his much earlier offering from 1958, *They Live in the Sky*. Trevor began filming in the Mojave Desert in the late 1950s and found he could capture UFOs in their invisible state using just a standard camera loaded with infrared sensitive film. He also discovered that many of these invisible UFOs were alive, having their form expressed in a heat-based existence which many people often refer to as plasmatic.

This interested me and I eventually went on to get a new digital camera in December 2009, which I then sent off to a company in the USA for an infrared conversion. During the process, the hot mirror was taken out and replaced with a 720nm infrared pass filter, which now blocked the visible light spectrum but allowed the invisible infrared radiation to pass through to the camera's sensors.

The infrared part of the spectrum appears just before the red of our visible light spectrum, and the Latin term *infra* literally means "before." At the opposite end of the visible light spectrum we have the colour violet, and just beyond the violet we have ultraviolet. Both infrared and ultraviolet light are invisible to normal human sight, but by using a converted camera or camcorder we are able to probe and explore these hidden realities.

When I first began filming in the infrared it was still freezing cold and the weather didn't seem like it was ever going to warm up. Ice and snow still covered much of Oxfordshire, and for a few months I just took random shots of the sky while getting myself used to the camera. It wasn't until I began using an attraction method that I started to get my first results, and throughout the summer of 2010 I began to capture quite a few different invisible forms. During 2011 I began filming in the ultraviolet part of the spectrum using a full spectrum Sony Handycam which I fitted with an external ultraviolet pass filter, and very soon the inhabitants of this invisible realm began to make themselves known to me.

I eventually put together a Quest for the Invisibles website, and made a few appearances on BBC Radio Oxford to talk about my work, as well as making a series of short videos for my Quest for the Invisibles YouTube channel. Over the next five years I went on to capture many more bits of footage in both the infrared and the ultraviolet, revealing an array of different invisible life forms, as well as various craft-like objects that also appeared to be present in our atmosphere.

Quest for the Invisibles documents my work from early 2010 up to late 2015, so join me on a journey into the invisible yet corporeal world that borders our own physical reality, and learn how you too can capture your own footage of invisible UFOs. Anybody can record this type of footage using readily available standard equipment, and ultimately, all it takes is a little time and patience.

CHAPTER ONE

A New Reality

Christmas day evening, 2008, started out just like any other evening with my usual dinner at around six o'clock, followed by the first of two nightly dog walks. I always look forward to going out with the dogs especially around Christmas time as it is so much quieter, and the little lane in which I live is almost devoid of traffic. Due to an ongoing problem with my right knee and lower back, I had recently been finding it difficult to walk both of the dogs together, so for the last week or two I started to take them out separately. Being terriers, they are both full of energy and ready to chase anything that moves. This is okay when you have full mobility, but with my limited back movement at that time, even the shortest walk could sometimes be hard work.

My youngest dog Alfie was only six months old and a half hour walk was about all he could manage before begging me to pick him up and carry him home, so I would take him out first and then go out for a much longer walk with my older dog, Lucy. As the temperature had started to drop dramatically over the course of that evening, I decided to put Alfie's coat on before we ventured outside, so after another ten minutes of fumbling around we set off up the driveway and out along the lane in the direction of the local park at around 6:45 pm.

This was a route that I had been taking for many years with nothing strange ever occurring, but unknown to me at the time, this night was going to be very different from any previous dog walk. From what would occur that evening, my life would never be the same again. This particular night a freezing fog covered most of the local area and although visibility at ground level wasn't too bad, many of the uppermost branches on the taller trees were obscured by a much thicker band of fog that appeared to be forming at around tree-top height. This thicker fog was much darker and

A New Reality

denser than the ground fog and an almost perfect line could be seen in the sky where the two met.

Within about twenty minutes we were at the local park which consisted of a football pitch, changing rooms, and a few assorted swings, slides and a roundabout. There didn't appear to be any other dogs in the area, so I let Alfie off the lead so he could have a good run around the park, while I sat on a swing and smoked a cigarette and pondered life. It didn't take long before the cold started to get the better of me, so after about ten minutes I put Alfie back on the lead again, and we began our walk home.

After leaving the park we followed the main road for a while, before turning off and heading across the grass towards a small electrical substation which had a small alleyway running alongside it. This alleyway was a handy shortcut home and was often used when out walking the dogs, as it came out just up the road from where I live. Upon reaching the substation we walked up the alleyway, pausing every now and then for Alfie to have his usual sniff around. Alfie, however, didn't seem his usual self and something appeared to be bothering him. He kept stopping and looking around as if distracted by something or someone. This was unlike his normal behaviour as he is quite a laid back dog. He doesn't notice much going on around him, especially when he is busy sniffing around. But something was clearly affecting him and it made me uneasy.

We went further along the alleyway until we were only a few feet from the end, when Alfie turned around and began pulling on the lead, trying to go back the way we had come. I had never seen him act like this before, so out of curiosity I let him have his way and we walked back towards the substation. After a few minutes of looking around and finding and hearing nothing, I came to the conclusion that he was probably picking up the scent of another dog or perhaps one of the local cats, so decided it was time to head home. I bent down to pick him up, ready to carry him the short distance there. As I came up with him in my arms, I happened to be looking up at the area above the substation and all of a sudden my attention was drawn to a faint glow that appeared to be emanating from deep within the fog. It appeared to be a light of some sort, but was still quite far off in the distance — maybe two or three hundred feet away at the time; so I wasn't exactly sure what it was.

At first I thought it was probably a helicopter or an aircraft that was off in the distance, flying into the Dalton Barracks Army Base, which is only a few miles outside of Abingdon. I often see Hercules aircraft and various Chinook helicopters flying backwards and forwards over the area whilst out walking the dogs. They tend to be quite noisy as they pass overhead, and always have regular flashing lights. This object, however, was completely silent as it moved slowly through the thick fog, and another thing that was evident was that this object was at an extremely low altitude, which I calculated to be only about a hundred and fifty feet, so that cancelled out the plane and helicopter options.

My next thought was that it must be a Chinese lantern that had probably been released by someone from a nearby garden as part of their Christmas day celebrations, as the skies had been full of them over the past few weeks. As it approached ever nearer the brightness became more and more intense and I could clearly see that it was a totally different shape to any of the Chinese lanterns I had ever seen. Another big difference was the size of the object, as I could see that it was way too large to be a Chinese lantern, and there was no sign of a flame or any characteristics normally associated with one.

Every so often the glow would grow fainter and would almost disappear as it moved slowly through the thicker fog, only to become brighter again as it passed through a thinner pocket. I stood there wondering what on earth was heading my way; it was a very strange sensation and I found the whole thing quite unnerving. Although part of me was really excited, the other half of me was erring on the side of caution, as I began to suspect that this could be something strange. In all my years of living in the area I had never seen anything flying so low, especially through thick fog and so close to the tops of the houses below.

Still, the glowing object continued to move closer until it was around fifty feet away, and again it disappeared into a another thick pocket of fog and was lost from sight for a few seconds. I

A New Reality

could see absolutely no sign of it at all, nor could I hear anything, but I calculated that it would probably pass right over the substation if it stayed on the course it appeared to be following. I decided to walk a little way back up the alleyway with Alfie to get myself into a better position to observe whatever was heading my way. After what seemed like an eternity I began to see a faint red glow that grew in intensity; it gradually became brighter until the object finally emerged from the fog, passing directly above me in all its glory.

I had a clear view for a minute or so as the object passed right overhead. I stood there totally stunned and almost frozen to the spot, staring up at the sky in total disbelief, trying to take it all in. I have never been so in awe of anything before in my life and it was a very moving experience — one that I will never forget. I estimated that the UFO, for want of a better word, was roughly the size of a large transit van. It moved at little more than walking pace and was completely silent. I found it absolutely astounding to see something like this at all, let alone whilst out dog walking, and at such a low altitude as well. I knew then and there that this was no normal craft in any sense of the word; it looked like nothing I had ever seen before and I was truly in a state of shock.

As you will see later in the photograph section of this book, the object was not the classic saucer shape. It looked more compact and was very smooth and rounded in appearance. My first impression was that it looked very much like a child's spinning top, with the curvature of the main body and the dome-shaped top and bottom being clearly visible. As it moved slowly through the dark atmosphere it slowly pulsated with a warm glow that gradually became brighter, going from red through orange and scarlet. At times it reminded me of those deep-sea creatures that you see on nature documentaries that live in the darkness of the deep ocean, the ones that have the means to produce luminescence.

The glow it gave off seemed to go through a regular cycle of colour change and it became more intense at the end of each cycle. I figured that perhaps this was something to do with its propulsion system. The whole thing appeared to be made of light, with no sign of any external features such as windows, markings or landing lights — things that are a standard feature of a conventional aircraft. I began to wonder if perhaps it was an alien craft or a surveillance drone of some sort. The UFO appeared to be moving under intelligent control as it kept the same speed and altitude, and this allowed it to remain hidden within the thick band of fog and made it almost impossible to see.

After passing directly overhead it continued on a perfectly straight course taking it over a nearby housing estate, and I estimated that it must have been about a hundred feet or so above the rooftops of the houses. I couldn't believe what I was seeing, and even remember pinching myself a couple of times to make sure I wasn't dreaming — that's how surreal the experience was for me. It's hard to put it into words and describe exactly how I felt, but there was a mixture of both fear and excitement combined with total disbelief at the situation.

Still in shock, I reached into my coat pocket for my new digital camera, desperate to capture a photograph of the object. After a few seconds of fumbling around I suddenly had the sickening realisation that the camera was in the pocket of my other coat, and not the one I had on. I was absolutely gutted; I just couldn't believe it. "Of all the times to not have a camera on me," I kept thinking. The annoying thing was that earlier that day I had been walking around filming anything and everything with it, and it barely left my hand for most of the day. Figuring I wouldn't need it when I walked the dogs later that evening, I had left it in my fleece which was at home, hanging on the coat rack, and that was at least twenty minutes away.

I found it hard to think straight and started to panic a little. I just didn't know what to do — whether to follow it or get home and grab the camera. Ultimately, I decided to keep following it and take in as much detail as I could before it had a chance to disappear on me. As the UFO began to head diagonally across the housing estate it was totally enveloped by the fog once more. Since I could no longer follow directly beneath it, I chose to run up a nearby road in order to get in front of it so I could watch it pass overhead again. As I ran with Alfie past the rows of houses, I had to stop myself from knocking on people's doors and shouting out "UFO ahoy" at

A New Reality

the top of my voice to make them aware of what was flying just above their rooftops. Had it not been Christmas day evening I might well have started doing that.

My heart was beating fast and my mouth was completely dry. I was also shaking from the adrenaline and finding it hard to comprehend what was going on, so I sat down with Alfie for a while against a fence and smoked a cigarette whilst I gathered my thoughts. I didn't know what to do and found it hard to think clearly. For a minute or so I considered ringing the police from my mobile phone and reporting it, but didn't think they would take me that seriously, especially with it being Christmas day evening. They would probably think I was drunk or on drugs or something, and besides, what could they really do?

After a while I decided to ring one of my close friends who lived just down the road, hoping he could drive up from his place and witness what I was seeing, but to my disappointment he was away in Norwich visiting another one of our friends who had recently moved there. I decided to continue following the UFO, which had now passed over the housing estate, still almost hidden by the thick fog, and heading toward the local park where I had previously been with Alfie. As it approached the park area it slowed down a little, then came to a stop right above a small forested area which had a path running right through it. I stood there on the path about a hundred feet away, just staring up at it in total shock and disbelief, wondering what might happen next.

After a few minutes I began to feel uneasy; all sorts of scenarios started running through my head, most of them fuelled from years of watching Star Trek, Star Gate, and other sci-fi programmes on television. I wondered if the object had a crew and if so, where did they come from? And more importantly, why were they here in Abingdon, and what did they want? If there was a crew piloting the UFO, then they might be aware of me standing there looking up at them. I must admit that I started to get nervous. I looked on my mobile phone and saw it was nearly eight o'clock. By now the park was almost covered by the fog and it was absolutely freezing.

I scanned the park, desperately looking to see if there was anybody else in the vicinity like another dog walker, or perhaps a

Quest for the Invisibles

jogger or two. It is rare for that park to be so empty, as most nights there would be at least two or three other dog walkers and a scattering of youths hanging around by the changing rooms. I wanted to see if somebody else was witnessing this, but the park was completely empty. It was just me and Alfie — and that made me very anxious.

I decided the best thing to do was rush back home and grab my camera as fast as I could, and then try and get some photographs of this strange glowing object before it disappeared. I picked Alfie up and we made our way home. I kept turning around and looking back to reassure myself that it was real and most importantly, that is was still there and not starting to move off. I arrived home in about fifteen minutes by half walking and half running.

By the time I got home and dropped Alfie off I was still in shock and out of breath; I had a few cuts and bruises and rips in my trousers where I had fallen a few times on the run back, but the adrenaline had kept me going. It took a while to get my breath back. After an extremely quick cup of tea, I grabbed my other dog Lucy, along with my camera, and we raced back towards the park as fast as our legs could carry us. Lucy was picking up on my excited state and eagerly ran alongside me, although it wasn't long before I returned to a walking pace as I wasn't in the greatest shape at that time.

After reaching the edge of the park, I let Lucy off the lead so she could have a run around, and I cautiously approached the area where I had previously stood looking up at the object. There was no sign of it, not even the faintest hint of a red glow coming through the fog. My heart sank and I felt totally despondent, staring up at the grey mass of clouds and fog. I whistled for Lucy to come, and after putting her back on the lead we walked around the park a few times to see if I could spot anything. I was hoping that the object would appear again from the thickening fog, so I could snap a photograph or two for posterity.

I continued looking for the object for another twenty minutes, but became so cold that I decided to go home, drop Lucy off, grab a quick coffee and go out again one more time. On the walk home I laughed as I thought about all the people sitting happily in their

A New Reality

front rooms and watching Christmas day evening television, totally unaware of this strange craft-like object passing just above their roofs; I often wonder if it caused any type of interference to their televisions.

After dropping Lucy off and grabbing that quick cup of coffee, I ventured out once again into the freezing night, still desperate to capture a photograph or movie mode footage of the object. This time I decided to walk around my garden first and see if I could catch a glimpse from there, as one side of my garden faced the general direction of the park. I was totally worn out by this time, not to mention cold; I could hardly feel my fingers, but I continued for a further ten minutes, staring into the distance and scanning the horizon, whilst mentally preparing myself for yet another trip back down to the park.

At first I could see no sign of any type of light in the sky, but I kept on looking as I was determined to see the UFO again — but all I could see was the swirling mass of grey and black, which was the sky. I stood there trying to determine if I should return to the park, when suddenly I saw a flash of light way in the distance. It seemed a lot further away than I had anticipated, so I was unsure at first if this was the same object or not. A couple of minutes later another flash of light caught my attention, then a few seconds later another, then another. I began to realise that this was the light from the object, penetrating through the fog as it became visible once more.

I pushed my way through the trees and thick undergrowth at the rear of my garden trying to find the best position to take a photograph from. Just as I reached into my coat pocket for the camera, another flash of light caught my attention. This time I could clearly see that it was the same object, although at a much higher altitude than it had been and was a few hundred yards from its original static position above the park. I turned the camera on and tried to put it on the automatic setting for photos, but due to the lack of light and my unfamiliarity with the camera, I soon gave up and set the camera to movie mode. I figured that would give me a greater chance of catching something, as it was taking several shots per second, rather than just single shots.

Quest for the Invisibles

I anxiously aimed the camera, trying to get the object in my viewfinder, but could hardly see it and kept losing it. My heart was beating so fast and I was shaking like a leaf as I struggled in the darkness with limited visibility. Every now and then I lost my footing and slipped down the bank of the old riverbed at the end of my garden, and on several occasions I actually dropped the camera and had to feel around on the ground trying to locate it.

It is all quite laughable now, but deep down I wish I had been more prepared, but that's the way it goes. The hardest part was trying to focus in on the UFO, as my line of sight was almost entirely blocked by the large bare tree branches that dominated the area in front of me. I filmed as best I could for another ten minutes, until finally the UFO disappeared into the thick mist and I lost sight of it for good. I decided to head back indoors. I was completely drained and it was getting late; I knew I had done my best under rather difficult circumstances and I was pretty confident that I must have caught at least something on camera.

The next morning, which was Boxing Day, saw me up early listening to all the local radio stations expecting to hear how a UFO had gone over a part of Abingdon at very low altitude and been seen by loads of people, but I didn't hear one single bit of news on the subject. I was totally flabbergasted and found it hard to believe, so I checked the Internet, with no luck, to see if there was anything about the incident. I also read a few of the local papers but found no mention of it. I can't believe that I was the only person in the area to have seen the UFO, and still wonder if I should have knocked on people's doors and brought their attention to this once in a lifetime event.

The more I think about it though, the more I realise that the UFO was only visible as it moved through the thinner pockets of fog, and remained far less visible while in the midst of the thicker fog. On several occasions I lost visual contact with it for a few minutes. The fact that it was Christmas day evening eliminated many potential witnesses to the incident, and if my memory serves me right, I can recall seeing only a couple of cars driving along the main road into town. This is a big contrast to a normal evening where I would sometimes bump into at least half a dozen different

A New Reality

dogs and their owners, as well as having to wait for the endless lines of traffic to clear so I could actually cross the road.

The fact that I was in the right place at the right time was due to my dog Alfie; if it were not for him pulling on his lead and insisting we walk back up the alleyway, I would have completely missed being a witness to this strange occurrence. I still don't know to this day what was affecting him that night and causing his strange behaviour, although I suspect he may have sensed the UFO long before it came into my visual range. My main regret is that I didn't have my camera when I first saw the object, as I was in the perfect position to take dozens of clear shots or even movie mode footage as it passed overhead. Since that evening I have made it a rule to never leave the house without a camera in my pocket.

After uploading the half dozen pieces of movie footage on to my computer the next day, I was shocked to discover that I had only managed to capture one frame that showed a good image of the UFO. Everything else I had taken that night turned out to be extremely shaky and blurred and therefore no good. After enlarging the good image and giving it a slight emboss, the curves and rounded edges of the body stood out more and it looked exactly as I remembered it. Just seeing it again made the hair stand up on the back of my neck, such was the effect that the whole experience had on me that night. I was relieved that I had managed to capture one image that confirmed what I saw, and although I felt pleased with myself, the encounter had raised more questions than it had answered.

CHAPTER TWO

The Search for Answers

For months following the encounter I could think of little else from the moment I awoke until the moment I fell asleep; I would spend hours on my computer watching as many UFO documentaries as I could possibly find on the Internet, as well as lots of footage recorded by people on their own cameras and camcorders. I found the sheer amount of available footage to be overwhelming, and although many of the clips I viewed were somewhat dubious, there were many more that were convincing and appeared genuine enough to warrant further investigation.

During the I time spent trawling through the mass of YouTube footage, I didn't happen to come across anything that looked remotely similar to what I had seen that night. The whole situation was frustrating, but my curiosity had gotten the better of me and I wanted to know more. I could not comprehend carrying on with my life as if nothing had happened, and I needed more answers. Those answers were not going to come from watching videos of other people's footage, and I eventually realised that I was going to have to start my own investigation into the UFO phenomenon.

By now it was June 2009 and it felt as though time was slipping away — six months had now passed since I saw the UFO that Christmas Day evening, and I felt no closer to finding answers than I had been on the night of the encounter. I knew I had a hard task ahead of me, but the whole experience had affected me so much and left so many unanswered questions in my mind that I felt driven and compelled like never before to try and gather more photographic evidence to prove the existence of UFOs, if only for my own records, and I now felt I was on some kind of mission.

As I began telling a few friends and family members of my intentions to start filming UFOs, it began to dawn on me that I knew absolutely nothing about cameras or photography. The only

The Search for Answers

reason I had a camera in the first place was because my brother-in-law got me one for Christmas, and if it wasn't for his generosity I would likely still not own one. My only experience with a camera prior to receiving one for Christmas was taking the odd holiday snapshot with my parent's little point and shoot Kodak film camera, and that was as simple a device as a person could possibly use.

The more I thought about it the more I began to realise the magnitude of the task at hand, and I began to wonder if I would be better off just getting on with my life and forgetting about the whole thing. Another realisation that was also weighing heavily on my mind was how and where I was going to locate the UFOs to photograph in the first place, even if I did get myself a better camera or perhaps a camcorder. I considered myself very fortunate to have seen the Christmas day UFO as it passed right overhead, but I knew the odds of that happening again were remote, to say the least.

These thoughts continued for several months and I went through a whole array of emotions as I tried to come to terms with what I had witnessed that night. I had to get more evidence, but didn't know where to start. I began to feel there was nobody who could provide me with the answers I needed or even guide me with my photographic quest. Although I had quite a few books and magazines about the subject at home, none of them seemed to cover any techniques for the actual filming of them.

Just as I was at the point of despair and starting to consider giving up on it all, a good friend happened to visit one night and he too had an interest in UFOs. He excitedly told me of a documentary type film he had seen recently on the Internet entitled *UFO – The Greatest Story Ever Denied*, so a few days later I managed to track down this film produced by Jose Escamilla, and I spent the next week watching it over and over again, such was the impression it had on me. I still consider it to be the most important and influential film I have ever seen, and as a result of watching it, I discovered the pioneering work of a man named Trevor James Constable — and thus find the direction I needed to go in order to begin my quest to film UFOs.

Quest for the Invisibles

Part of the documentary described how Trevor James Constable discovered that many UFOs exist in a light spectrum invisible to the human eye, and how in the late 1950s he had managed to photograph them in this invisible state using a standard camera loaded with infrared sensitive film. It also mentioned that Trevor had found that many of these invisible UFOs were actually living organisms.

I couldn't believe what I was hearing and seeing and it completely blew me away. I had never given a thought to the idea that UFOs might exist in the invisible spectrums, let alone that they might be alive. I had always assumed that they were ships from other planets, which is the general hypothesis that is fed to us via the media, and it had never occurred to me that they could be anything other than this. Watching this film had opened up a new world of possibilities for me regarding UFO photography. I knew I had to find out more about Trevor James Constable and his work, and eagerly set about my research.

I found out that Trevor, who was a well-renowned author and historian, had written ten non-fiction books on famous airplane fighter aces, as well as serving 31 years at sea, 26 of them as a radio officer in the U.S. Merchant Marine. In 1958 he had another book published entitled *They Live in the Sky*, and this covered his pioneering work in the Mojave Desert of Southern California when he first began using infrared sensitive film to capture invisible UFOs. Another book followed in 1976 entitled *The Cosmic Pulse of Life*, and this covered Trevor's later photographic work and his experimentation with cloudbusters as an attraction method. The book had recently been updated and re-released by The Book Tree and was now on its fourth edition – packed with more photographs and information and well worth reading.

I was quite keen to get hold of this new, updated version of the book, so immediately ordered myself a copy, but Trevor's other book, *They Live in the Sky*, was quite rare and not easily found. After about a month I finally found a hardback copy for sale on Amazon. It was listed as being in average condition and slightly worn, but as the book was published in 1958 I figured most copies would be in a similar condition, so I purchased it.

The Search for Answers

Within three or four days of ordering a copy of *The Cosmic Pulse of Life,* it arrived. I excitedly opened the package and started to familiarize myself with the many photographs and the whole host of information contained within. I had never been so eager to read a book for as long as I can remember, and I spent every spare moment reading it. I will give a brief outline of Trevor's work in this chapter, but strongly advise that anyone interested in this subject should get themselves a copy of Constable's book *The Cosmic Pulse of Life*, as I feel it is essential reading for anyone wanting to photograph the invisible. The book will give you an understanding of the basic theory of living UFOs, as well as the techniques for objectifying and photographing them.

Trevor James Constable's introduction to UFOs came via the books of Major Donald Keyhoe, which were published in the 1950s. Trevor also met with George Van Tassel, an early pioneer in telepathic communication with UFO intelligences, and soon became a regular at Van Tassel's séance-like communications. They were held in a hollowed out chamber beneath Giant Rock in the high desert of Southern California. But it was the Nansai-Shoto incident mentioned in Keyhoe's book, *Flying Saucer Conspiracy*, that got Trevor thinking about capturing images of UFOs on film. The incident happened during the Pacific campaign against Japan, and was called the Nansai-Shoto incident due to its occurrence near the islands south of Okinawa. It is possibly one of the most significant of wartime detections, and Trevor talks about the incident in *The Cosmic Pulse of Life*.

It began when an aircraft carrier detected a huge force of assumed enemy aircraft approaching from the northeast. Initial radar contact was 120 miles, and the return radar echo was very large and the supposed enemy force was estimated to be as many as 200 to 300 strong. As they came within a 100-mile range their speed was determined to be nearly 700 miles per hour, and at that time there was no known aircraft in the world that could attain such a speed. Eventually the incoming force spread out into two formations from the main body, as though preparing to attack the two aircraft carriers that were patrolling the sea south of Okinawa.

Quest for the Invisibles

Only twelve fighter planes were available at the time due to the others being away on a mission attacking nearby Japanese positions; these twelve were hurriedly scrambled and sent to intercept the radar-identified attacking force. The American fighter pilots had unrestricted visibility due to bright weather and at 15,000 feet they could see for about fifty miles. Despite being directed accurately from the carrier to intercept the "enemy," these veteran Navy fighter pilots could see no sign of the attacking force, even when they were directly above them as shown on the ship's radar. The UFOs were totally invisible to the twelve experienced Navy pilots despite the fact that they had unlimited visibility that day, as well as having the presence of many trained observers aboard the ships — who saw nothing throughout the entire episode.

Along with the Nansai-Shoto incident and other similar incidents where UFOs were detected by radar but not visible to the human eye, there were also many accounts where visible UFOs had simply disappeared and appeared again whilst under observation. Trevor decided to concentrate his attention on these particular types of sightings where UFOs had materialised from the invisible and dematerialised back into the invisible. Realising that the infrared portion of the spectrum adjoins the microwave spectrum, which is the portion of the spectrum in which radar operates, he figured that perhaps the infrared was the place to start looking for these invisible UFOs.

Armed with information gained from certain etheric beings with whom he had telepathic communication with for a brief time, and also with the wise words of the accomplished occultist Franklin Thomas, Trevor set off with his good friend Dr. James Woods to the Mojave Desert with the aim to capture invisible UFOs on infrared film. He soon realised that the one problem they faced was how to attract the UFOs that they sought to photograph. The answer was provided by Franklin Thomas, who suggested that he perform cyclical repetitions of the "Star Exercise," an esoteric procedure by which the human body force is strongly energised.

Performing the Star Exercise in a cyclical manner such as this will result in a regular pattern of bio-energetic pulsations in the ether, and these pulsations, it was found, would eventually attract

The Search for Answers

the various UFOs to the immediate area, where they could then be photographed. Trevor decided he would act as the "bio-energetic" beacon and Dr. Woods the photographer, although Trevor would also have a camera loaded with infrared sensitive film at the ready, should anything be detected. The full procedure for performing the Star Exercise can be found in *The Cosmic Pulse of Life,* along with the technique for objectifying invisible UFOs.

In the summer of 1957 they began driving to the Mojave Desert at night, where they would then set up camp so they could be up at first light and ready to start the attraction process. Despite going back to the same place time and again and performing their attraction technique, they didn't seem to be getting the results they wanted. All that was about to change though, as on the morning of August 25th 1957 while Trevor and Dr. Woods were sitting outside having breakfast at their desert campsite, Trevor suddenly became aware of a presence above him. He described the impression he felt as overpowering, and he instantly sprang to his feet and began scanning the area immediately above. Against the clear blue sky Trevor could "see" a strong pulsation, and what he describes as a "shimmering variation to the otherwise smooth blue sky background." Dr. Woods was unable to "see" the object which Trevor described, so Trevor grabbed an already loaded Leica 35mm camera loaded with infrared film and fitted with an 870nm infrared pass filter, and began taking photographs whilst describing to Dr. Woods how the pulsation was moving right above them.

The first photograph taken by Trevor was when the pulsation was directly overhead, but as it moved toward the south and began to lose altitude, he was then able to frame portions of the desert terrain in the photographs along with the pulsation. Of course the biggest shock to come was when the photographs were developed and they finally got to see just what they had managed to capture on infrared film. They eagerly picked up the photographs from the developers and to their amazement they hadn't captured a ship from another planet as expected, but a huge invisible living organism very similar to an amoeba, complete with nuclei and vacuoles.

Quest for the Invisibles

All in all, Trevor took six exposures of the object(s) and most of these pioneering photographs appear in *They Live in the Sky,* and *The Cosmic Pulse of Life*. The last photograph taken that morning featured not only one of these "amoebas," but also another similar creature alongside it. Both creatures appeared to have eyes, and the other creature has an almost reptilian looking face, with what looks rather like a bill or beak beneath the eyes. In Trevor's earlier book *They Live in the Sky* the same photograph was labelled as "The Peekers" and is one of my personal favourites. Trevor went on to photograph many types of biological UFOs, or "critters" as he came to call them, and many of them were enormous in size.

By performing the Star Exercise in such a remote location as the Mojave Desert, Trevor and Dr. Woods managed to attract and photograph a whole host of invisible UFOs of many different shapes and sizes, as well as various strange force fields and auras. In one particular infrared photograph taken at 5:30 am in the Mojave Desert on April 26th 1958, Trevor can be seen in what he describes as a "UFO shower." In this photograph, taken by Dr. Woods, Trevor is standing there, silhouetted against the sky, with dozens of spherical, invisible UFOs all around him. Trevor also took a photograph of these same spherical UFOs at the same time as Dr. Woods was taking the photograph of him in the UFO shower. So both of them managed to objectify and photograph the same invisible objects at the same time, using two different cameras, independently of each other.

Trevor later became acquainted with the work of another pioneer of his time, Dr. Wilhelm Reich, via his daughter Eva. Working from the basic inventions of Dr. Reich, Trevor built a cloudbuster and on May 11th 1968, began using it in place of the Star Exercise as the primary attraction method. Using the cloudbuster to excite the atmosphere locally, they managed to attract and film more and more examples of invisible biological UFOs.

Trevor would later go solo with his work and went on to rent an office in North Hollywood where he continued to perform the attraction technique, photographed the UFOs, and developed the infrared film. He managed to photograph a huge range of different invisible forms, from flying discs to ovoid shaped biological UFOs,

The Search for Answers

many of them right over the business district of North Hollywood and all totally invisible to the human eye. Many of these great photographs are featured in the photo section of *The Cosmic Pulse of Life* along with a few other photographs from other reliable sources, including the Italian UFO research group GRCU (Gruppo di Ricerche Clipeologiche Ufologiche) who were based in Arenzano, Italy.

Although in his early photographic work Trevor often used an 870nm infrared pass filter in conjunction with high-speed infrared film, he sometimes used the infrared film without a filter, with equally good results. He later started experimenting with a Wratten 18A filter, which is primarily an ultraviolet pass filter, but also passes a decent amount of infrared as well, whilst blocking all visible light. In 1975 he found that by using an 18A filter in conjunction with a low-light Super 8mm movie camera loaded with Ektacolour 160 colour film, he could shoot from an airline window and capture UFOs in full colour as they flew alongside the airliner at high altitude. The rest of the passengers on these flights were totally oblivious to what was going on immediately outside their aircraft, which was probably a good thing, considering what Trevor managed to capture on film.

He describes how once, while filming from an airline window at two frames per second with the movie camera, he managed to take a chain of twenty-five or so exposures that showed a yellow flying disc accompanying the airliner. The UFO approached the airliner on a couple of occasions as if carrying out some sort of inspection. Trevor went on to film many types of strange, invisible UFOs from the airline window, and not all of them were craft by any means. He goes on to mention one exposure showing what he describes as a monstrous form with awful looking spikes on its back and what looked like a beaked head, and like the other UFOs, this was pacing the airliner at around 500 mph.

The fact that Trevor managed to get these photographs despite using an 18A optical filter which is designed to absorb all visible light and colour, challenges the accepted thinking on the origin of colour. It was obvious to Trevor that these colours were manufactured between the lens and camera filter, and he called this mingling of the two invisible ends of the spectrum the reverse or dark spectrum. Using this technique creates an artificial darkness during daylight, and the UFOs "reverse out" from this artificial darkness and then inscribe themselves onto the film's emulsions.

Trevor finally gave up his photographic pursuit of UFOs in 1979 to concentrate on weather engineering, but shortly after giving up UFO photography he happened to receive a letter from Luciano Boccone, who was president of the Italian UFO research group, the GRCU (Gruppo di Ricerca Clipeologiche ed Ufologiche). The group was made up of technicians and engineers who were drawn together by their interest in the UFO phenomenon.

The GRCU had been researching the UFO phenomenon since 1976, using infrared film and a range of detection instruments including temperature indicators, Geiger counters, ultraviolet detectors, and precision magnetic compasses. They were totally unaware of Trevor's work in a similar vein twenty years earlier, until they later read of his work in an Italian translation of Brad Steiger's book, *Gods of Aquarius*. They were very keen to share their findings with him, and Trevor began working with the group on a consultative basis.

The GRCU managed to take some incredible infrared photographs, and many of them are featured in Luciano Boccone's book, *UFO: La Realta Nascosta* (*UFO: The Hidden Reality*), which was published in 1980 in Italian only, as far as I am aware. The book features infrared photographs from three different sources, as well as a selection of photographs taken from *The Cosmic Pulse of Life*. The photographs contained within cover a whole range of invisible phenomena, from haloed human-like entities and strange griffon-like flying creatures, right through to luminous spheres and other plasmatic life forms, both at low and high altitude. A majority of the photographs taken by the GRCU were as a result of instrumental detection, and this proved to be a very reliable method over the years.

I still find it hard to believe that in all my years of reading UFO literature and watching countless documentaries about the subject, I had never heard the name Trevor James Constable, let alone

The Search for Answers

come across any of his work. It is apparent that Trevor's findings upset a lot of people and went against the conventional "ships from other planets" theory, much to the dismay of many other UFO researchers and organisations, many of whom were still sold on the idea that UFOs were interplanetary vehicles and nothing else. The mechanistic thinker, it seems, has a great problem facing the reality that some UFOs could be alive, and Trevor had his fair share of being treated appallingly by people who really should have known better.

So having now become associated with the work of Trevor James Constable, I felt I was starting to understand a lot more about the UFO phenomenon. Much of what I read in Trevor's books made perfect sense to me and certainly changed the way I thought about many things, including my own life and this physical reality. One thing is for sure, my encounter with the glowing UFO on that freezing cold December night had been a life changing experience for me, and my subsequent interest in infrared UFO photography was starting to become a matter of urgency as far as I was concerned. I hadn't felt this keen to start a project for some time, but after reading about Trevor's adventures in the Mojave Desert and seeing all those amazing infrared UFO photographs he managed to take, the world suddenly began to seem a lot more interesting again and I felt a new lease of life.

CHAPTER THREE

Applying the New Knowledge

I spent most of summer 2009 continually reading *The Cosmic Pulse of Life* as well as *They Live in the Sky,* as I was determined to familiarise myself with the huge amount of interesting information contained within. Over the years I have read many books and articles on the UFO subject and it is something that has always interested me, but I have always come away feeling slightly let down and feeling that I was no closer to discovering the truth. After reading Trevor's books though, I felt I was getting much closer to the truth, and they also gave me hope that I too would be able to take photographs of invisible UFOs.

I later managed to track down a copy of Luciano Boccone's book from a vendor based in New York. The book, which is in excellent condition and A4 in size, is packed with infrared UFO photos and information of all sorts. Although it was costly and is written in Italian, the pictures are worth more than words can say. I knew I had to get a copy as I kept coming across references to it. I also heard there were only a limited number of them printed back in the early eighties, and had not seen any copies for sale up until this point.

Luckily for me, I happened to know an Italian lady who is a friend of my family. She spoke both English and Italian and subsequently helped me to translate a small portion of the book. Getting hold of this rare book only served to increase my enthusiasm for all things invisible, and it also opened me up to the world of Geiger counters and other precision measuring devices with which the GRCU group managed to objectify most of their invisible UFOs with, prior to the actual photographic process.

As summer turned to autumn I started thinking about getting the equipment needed to start my quest. My initial plan was to go out and get myself a simple camera, perhaps even a cheap second

Applying the New Knowledge

hand one as suggested by Trevor in his book, and then try and find a good source of high-speed infrared film. I already had a dark room kit at home which had all the equipment needed to develop and print my own photographs, but after weighing all the options I put the infrared film idea on hold and looked into getting a digital camera instead. This decision was made due to the cost and limited availability of infrared film in the UK at the time, as the black and white high speed infrared film favoured by Trevor in his work was no longer being produced by Kodak due to a drop in demand.

I did, however, come across a few retailers who claimed to have large stocks of infrared film bought from Kodak prior to its discontinuation, but I couldn't be sure of its quality by the time it would arrive in the UK. After all, infrared film is so sensitive to light that it has to be stored and loaded onto the camera in complete darkness. Trevor and Dr. Woods found that whilst they were spooling their infrared film into cassettes from the 100 foot roles, that even the smallest pinhole of light hitting the film would cause fogging and make it unusable. This was another reason that put me off infrared film, combined with the fact that it could turn out to be so time consuming.

Although I felt totally driven towards the digital camera option, I wasn't sure if I would be able to use a digital camera to record in the infrared. I knew from watching YouTube videos that there were a few people said to be filming UFOs with infrared converted camcorders, but I hadn't come across any actual infrared camera footage. I didn't know at the time that you could get a digital camera converted, until I found an online gallery of digital infrared photographs of old buildings and scenery whilst browsing the Internet one day. The buildings looked very eerie, along with the dark sky and the white grass and foliage; it also looked very dream-like and just the sort of environment in which I could imagine some strange invisible creatures flying around.

I knew from reading Trevor's books that the infrared film he used often reacted to the energy of these invisible UFOs in strange ways. Many of the objects were recorded on the infrared film because their particular type of energy nullified the film's emulsions rather than reacting with them, and this gave a rather unique quality to the photographs and in many cases revealed a lot of detail. I didn't know if an infrared digital camera could achieve this, or even if invisible UFOs would show up in any great detail on the subsequent photographs. I had read plenty of articles that mentioned just how sensitive digital cameras were nowadays, so I decided to take the chance. Just as Trevor experimented with infrared film, I too would have to experiment, but with an infrared digital camera.

It was around this time that my sister had been looking to buy a new camera after her old one finally gave out. She decided to go digital and ended up buying a Canon G7; I immediately homed in on it, as it looked so easy to operate and took great photographs. After spending a little time getting acquainted with its various functions, I began to feel comfortable using it and started to fancy something along those lines. I did some research and discovered that the Canon G7 had been superseded by a newer version, the Canon G10. I read some good reviews about it, especially the quality of the photographs, and also its movie mode footage which records at thirty frames per second. After a couple of days, I decided this was the camera for me.

As Christmas 2009 was fast approaching, my family chipped in and helped me get the camera as an early Christmas present. By the second week of December, my new Canon G10 digital camera arrived. I studied the instructions and watched a couple of YouTube videos explaining all the settings. I was really excited and had moved a step closer to filming in the infrared.

The next decision confronting me was how and where I would get this camera converted so I could continue my quest. Over the previous few months I had been doing a lot of research on both infrared and full spectrum conversions, as well as watching more videos about infrared photography. There were a couple of names that kept cropping up. Both of these companies were based in the USA and seemed to have good track records and great websites. The two companies were Life Pixel and MaxMax, and both offered a couple of different digital camera conversion options. MaxMax also offered a camcorder conversion service and had a large range of external filters available.

Applying the New Knowledge

In my case, the two options available were either an infrared conversion or a full spectrum conversion. The infrared conversion involves removing what is referred to as the hot mirror and replacing it with an infrared pass filter. The hot mirror is a filter that blocks the infrared and a good majority of the ultraviolet, i.e. light from invisible sources, which would otherwise discolour the photographs with unnatural colours. Once the hot mirror is removed and replaced with an infrared pass filter the camera can now see in the infrared, while the visible spectrum as well as the ultraviolet is now blocked. With the full spectrum conversion the hot mirror is again removed but this time replaced with a piece of clear quartz, meaning that the camera can now see the full spectrum, i.e. ultraviolet + visible light + infrared. External filters can then be used in combination with the camcorder to pass or block the desired part of the spectrum.

After much deliberation I finally decided to send my camera to Life Pixel for their standard infrared conversion, and this was mainly due to a friend of mine who had said what a good job they had done on his camera. With the standard conversion the hot mirror is replaced with a 720nm infrared pass filter, so everything under 720nm is blocked and only the invisible infrared radiation is allowed through to the camera's sensors.

I sent the camera off for conversion in late January 2010, and by the second week of February it had arrived back fully converted and ready for me to start filming. At long last I had my "eye into the invisible world." I couldn't believe I had actually done it and although keen to make a start, I still felt reluctant to go outside as there was still snow and ice everywhere—so I kept checking the weather forecast on a daily basis.

The temperature here in Oxfordshire and probably most of the country hadn't changed for what seemed like months. It hadn't gone much above freezing during January and February, and it felt as if the winter was never going to be over. By early March the sun began to put in an appearance and its rays were starting to melt through the last remnants of the snow and ice. I was itching to start my quest to film invisible UFOs, and didn't intend to let anything stand in my way, let alone the British weather.

Quest for the Invisibles

I began taking walks along an old railway track bed that runs near my home. This was the former Great Western Railway branch line that used to run into a small terminus station in Abingdon town. The line and station were closed to passenger traffic in the early seventies, but the line had continued to be used by the MG car works for a few years to move their new cars out of Abingdon. The former track bed has since been concreted over and made into a walkway and cycle track, and there were several sections along the walk where you could branch off and wander into the countryside and join up with one of the many paths that ran alongside or near the river Thames. I would continue along the track up to where the path ended, as here the original ballast still lay in place and the track bed turns off in the direction of the mainline. Although it was tricky to walk over, there were no icy parts, and this was a great bonus when trying to take photographs.

The view through the LCD screen on my Canon G10 looked great; it was the first time I had ever seen the world through the lens of an infrared camera, so the white grass and foliage looked rather strange and took me a while to get used to. I had set the white balance using a patch of thick grass in my garden, although I had tried a few other things like a green wooden fence and also various shades of green-painted wood, but the grass seemed to give the best result. After watching a YouTube video about my camera, I chose to set it to program mode—this gave me control over the white balance, which allowed me to see the image without the red tint that often accompanies an infrared conversion. I decided to shoot in RAW mode so that I could have an uncompressed version of the photograph as well as the JPEG version which is also provided. That way, I could keep the RAW photograph untouched and filed away for future reference.

When I first began going out with my camera I would find a place where I had good all around visibility, and then took dozens of photos of the horizon, moving slightly each time in one direction. I continued this until I had turned three hundred and sixty degrees, then would repeat the process many times. This is how I began my venture into infrared photography. I would normally shoot for a couple of hours then head home and load all the photographs onto

Applying the New Knowledge

my computer. Later in the evening I would go through them all, looking for anything that seemed out of place.

I started noticing an odd, spherical-shaped glowing object appearing in my early snaps along with other strange shapes. These would usually appear on one photo out of a whole batch of photographs taken of the same area of sky. These objects looked to be far in the distance, perhaps a few thousand feet or more and almost out of range of my camera. I soon realised I would need a method of attraction to try and bring the UFOs nearer, so I began experimenting with the Star Exercise. Due to my ongoing back problem I found it hard to stand upright for any length of time. Another problem that hindered me was the sciatica that I often experienced, as this could get very painful indeed. It made it almost impossible to keep my concentration, especially combined with the pain killers I had to take on a daily basis.

I found it all very frustrating and I sure could have done with a Dr. Woods-type character to help me, but as there were none offering their services at the time I knew I was on my own and had to try another approach. After first reading about Reich's cloudbusters in *The Cosmic Pulse of Life* I developed an interest in them. I found the entire subject fascinating and began looking into the work of Dr. Reich and orgone energy. I soon thought about building my own cloudbuster to help with the attraction process, and although there weren't any plans openly floating around, I understood enough about the design and theory to construct one.

Dr. Reich's original cloudbuster was basically a collection of hollow metal tubes which were mounted on a pivoting device similar to a seesaw, thus enabling one end of the pipes to be elevated and aimed into the atmosphere at any elevation, whilst the other end was grounded in water. Dr. Reich theorized that the hollow metal pipes would draw orgone energy from the atmosphere into the water, thereby allowing the manipulation of the weather.

It was found that when the cloudbuster was accurately aimed at a random cloud, the device would eventually dissipate that cloud, making it disappear from sight. Having experimented with the cloudbuster in this manner, I can honestly say that it does work,

Quest for the Invisibles

and I have made several clouds disappear on many occasions. A by-product of using the cloudbuster is that it also attracts UFOs, both creatures and craft, and Trevor James Constable later used one regularly to attract and photograph UFOs in place of the Star Exercise.

I began experimenting with my designs for a medium-sized cloudbuster, and as luck would have it most of the parts were just lying around in my sister's garage, waiting to be recycled or thrown away. This saved me a lot of time and money and also gave me the chance to try out several different ideas. Towards the back of the garage I found a huge pile of hollow metal tubing. This had been the framework for a huge gazebo before the covering material had eventually become too brittle and torn to be of any use. These lengths of tubing were just what I needed, as each piece was slightly smaller than the last and they fitted together very well. These were used to form the draw pipes of the cloudbuster.

I completed my version of a Reich cloudbuster towards the end of March 2010, and straight away I began looking for somewhere to operate it from—somewhere away from the gaze of the general public. I eventually designated an area on some spare land towards the rear of my home, and this proved to be an ideal choice. It was almost totally surrounded by hedges and trees, and from here I had a good, all-around view above the treetops, which were mostly of average size. I figured this would be the ideal place to perform the attraction process from, as from the air the treetops would look like a green ring and perhaps become a focal point for any potential UFOs.

I began going out at the same time every day for around two or three hours, trying to get some sort of response from the invisible realm. I continued with this method of attraction for several weeks without any tangible results, but I was determined to keep at it. Unlike Trevor James Constable, I didn't have a desert area in which to perform my work, but soon realised I had to make the best of what I had, and that included the unpredictable British weather. I wondered if my location would hinder me in my photographic quest. After all, I was right in the middle of a large

Applying the New Knowledge

populated area and only about half a mile from the nearest town and its noisy traffic. I recalled, however, that Trevor had managed to photograph invisible UFOs in and around the Los Angeles industrial district, so figured I had a chance of capturing at least something.

I had no doubts about the capability of the cloudbuster to attract UFOs despite its simple design, and had nothing but the upmost respect for it. After just two weeks of operation the tubes were so magnetised that they had the power to spin a compass needle around, severely affecting its ability to determine true north. I had read similar claims made by Trevor in *The Cosmic Pulse of Life*, so when I originally found the tubes in the garage I tested their magnetism prior to construction and found they had no magnetism at all. Although I had not managed to photograph anything yet, I was convinced that I would be successful now that I had a means of attraction and a good infrared camera to film with.

By mid-April the weather was changing and the days were getting noticeably warmer and slightly longer. On the afternoon of April 16th 2010 it was a warm day and the sky was a beautiful shade of light blue, there were no clouds at all and visibility was good—an absolutely perfect day for filming. I had only been using my cloudbuster for a few weeks at this point, and was still trying to find the best technique for attracting "invisibles" to my location. I knew that the cloudbuster worked as Trevor's infrared photographs bore testament to this. But since he never detailed the actual techniques he had used, I was virtually making it up as I went along, having no guidelines to follow.

I decided to experiment with different elevations and aimed the cloudbuster pipes just above the top of the trees where I then performed slow, sweeping motions from left to right. After about fifteen minutes I stopped and began taking multiple photographs of the area just above the treetops. I then elevated the pipes by a few degrees each time so as to cover the entire visible horizon, and then continued taking photographs. After a few hours I was feeling a bit weary, so took another dozen or so shots of the area of sky where the cloudbuster was pointing, and decided to call it a day.

Quest for the Invisibles

After getting the images on to my computer later that evening, I viewed them one at a time, looking for anything out of the ordinary. I was impressed with the quality of photos taken by the Canon G10, and although I hadn't managed to capture anything so far, I still felt confident that I was on the right track. I finally got to the last three photos that were taken at the end of the afternoon's session. I viewed the first one then the second, and then flicked to the third and final photograph.

As I looked carefully at it I noticed a light coloured dome-shaped object, and it only appeared in the third of the three photographs which were taken one after another at $1/160^{th}$ sec. I zoomed into the anomaly and my heart nearly stopped, I couldn't believe what I was looking at, but sure enough—it had the classic UFO shape and looked like a craft of some sort. I took this as an omen to the future success of my infrared UFO photography. I had reached a milestone and this made me more determined than ever to keep on with my quest. As May 2010 came around the sun was getting higher in the sky and warmer days started to follow. I went out with my camera whenever I had a few spare hours, and at one point was filming three or four times a week. During mid-May 2010 the big news here in England and also much of Europe was the volcanic ash cloud that was drifting over from Iceland. As a result of this, all flights in and out of UK airports were cancelled and the sky was eerily quiet and free from the dozens of vapour trails that normally criss-crossed directly above my house.

On May 17th it was a sunny day and by mid-morning I was at my research area setting up my equipment. I had the whole sky to myself, and not a vapour trail in sight. The volcanic ash cloud was still drifting over Europe and causing chaos, leaving scores of people stranded on foreign soil. My sister was in Egypt with a friend and they too ended up having an extended holiday because of the situation. I was glad to be here in the UK, safe and sound.

After operating the cloudbuster for a while, I took overlapping photos of the horizon through the entire three hundred and sixty degrees. I started just above the treetops and gradually raised the camera slightly and started the whole process again, making sure that I covered as much of the sky as I could without missing any

Applying the New Knowledge

areas. I then repeated the process over and over again, all the time practising the method for "seeing" the invisible, which involves defocusing your eyes and looking past things rather than at them.

On several occasions I became aware of fast flashes of light in the area where the cloudbuster was aiming. I tried to react with the camera, but wasn't fast enough. At this point I wasn't sure if I was picking up on something invisible shooting across the sky, or if it was just my eyes reacting to reflections from the sun. I decided to focus my camera on the area of sky above a row of ewe trees that formed part of the visible horizon to the east. That was where I first became aware of the flashes, so continued taking dozens of photographs over the next hour before packing up for the day and heading back to the house.

After checking the day's footage I found over a dozen photographs that had luminous objects present. Some of them appeared to be v-shaped and many were almost pure white due to the reflected infrared radiation. Most appeared in just one or two photographs. It was impossible to tell how high up they were, but one thing was certain—they were way out of reach of my camera with its limited zoom. I experienced this many times and the problem was they didn't seem to want to come in any lower, and as I didn't have a decent zoom at the time, I was stuck.

I decided to purchase more memory cards for my camera as I had only been using the 4GB card that came with it. Unknown to me at the time, the 4GB card would allow me to take only three or four continual photographs before the camera needed to pause for a few seconds to process the information. This was because I was shooting in RAW mode, which produces bigger files than JPEGs and required much more memory. So I picked myself up a couple of 16GB cards; this seemed to do the trick, and it helped me shoot longer and take multiple shots without having to pause. In these early days it was all experimental on my part, as I had never read anything about digital UFO photography.

During these early months, as mentioned, I photographed many strange looking objects way up high but the images were far too small to see any detail. I began to wonder if I was ever going to get these objects to come down any closer so that I could start to get some clearer footage. At least when Trevor James Constable was filming in the Mojave Desert he had miles of space all around him and the isolation of the desert meant that the hustle and bustle of everyday life, with all its noises and EM transmissions, were but a distant dream. Any little disturbance to the ether in the Mojave Desert would have surely attracted a lot more than I ever could have hoped for here in the UK, working from the centre of a clump of trees in the middle of a built up area. This was the one big disadvantage I had compared with Constable, in that from his vantage point in the Mojave Desert he had a great view, especially from the top of Giant Rock where he could see the horizon on all sides.

In one photo, "UFO Ahoy!" taken by Dr. Woods in spring, 1958, in the Mojave Desert, Trevor is photographed standing close to an emergent UFO similar to the "ventla"-type craft described and drawn by George Van Tassel, and appearing on the following page in *The Cosmic Pulse of Life*. This cross-sectional drawing was originally published in George Van Tassel's *Proceedings: Issue 1,* in December 1953. In the photograph taken by Dr. Woods, the main pyramid-like body of the craft can be seen surrounded by what appears to be a whirling force field, which Trevor speculates to be the spinning energy field that forms part of the craft's propulsion system. At certain times he could feel the radiation as it beat against his own bio field. In parts of the photograph there are dark areas, which are suggestive of an area of high absorption of energy.

In some of Trevor's other infrared photographs there are also areas where no reaction had taken place with the film's emulsions, again leaving a dark black void on the film. A similar reaction took place on the series of photographs showing the "UFO shower," mentioned earlier, where many of the spherical objects appeared as almost completely black voids. A few, however, were more ovoid in shape and appeared to be expanding and contracting as they moved, similar to when we are breathing and our chest increases and then decreases in size with each breath.

These were fantastic photographs, especially knowing they were taken in 1958, not that long after the second World War.

Applying the New Knowledge

Photographs like these inspired me to get out there and get my own footage. The excitement of not knowing what the day's photographic catch will bring is exciting and addictive, and many other parts of my life faded by comparison as everything became enveloped by my infrared photography. The big question on my mind was whether I would be able to get the same detail in my footage as Trevor James Constable had in his photographs, and I began to wonder if I had made the right choice by going digital instead of using infrared film. I did some more research into digital cameras and became convinced that although they work in a different way to conventional cameras (that used film), they are still very sensitive.

I continued on with my work, trying to attract the UFOs with the cloudbuster and bring them nearer in the hope that I would eventually get some clear footage. I started recording in movie mode instead of taking random photographs. With the camera continually filming, I would be less likely to miss anything. I began setting the camera up on a tripod and aiming it exactly where the cloudbuster was pointing—above the tops of the trees. Despite my limitations, I could still zoom in a little to get closer to the action. Eventually my methods began to pay off, and by July 2010 I captured a few more invisible objects—this time at a much lower altitude. Many of them were spherical and appeared plasmatic in nature, while others appeared to be luminous forms of energy that changed shape as they moved through the atmosphere. I also captured a few objects that were pyramid-shaped or looked like discs, although these were a lot higher in the sky and moving at an incredible speed. It was obvious to me that the cloudbuster was doing the job and starting to attract "invisibles" to my location.

I continued recording in movie mode, and although I was now managing to capture things at a much lower altitude, I still hadn't filmed anything that looked like the UFOs that appeared in Trevor's books. Most of the things I had captured appeared pure white, and they seemed to lack any real detail. At least I was getting results at long last, and I knew the more I filmed, the more I stood a chance of capturing better footage. I continued experimenting with the cloudbuster and trying different techniques of attraction right up

Quest for the Invisibles

until the end of July. Then I took a break from my pursuit and went on holiday for a week to East Sussex, along with my sister and niece.

Although I enjoyed being away from home and in different surroundings, all I could think about was UFO photography, which was starting to become somewhat of an obsession. I knew that I needed to switch off from it all, so made a concerted effort to enjoy the holiday and forget about it for a while. On reflection I felt pleased with what I had achieved in the previous months, but knew I still had a long way to go and a lot more to learn about the invisible world. For now, I could relax and recharge my batteries. I still had the rest of the summer to continue on with my quest, and who knows what I would discover in the coming months.

CHAPTER FOUR

Skyfish

After returning from my week away in East Sussex, it was time to catch up on a few gardening jobs and other chores that I had been neglecting in favour of my infrared photography. When I returned to the photography it was already mid-August, and the UK was experiencing somewhat of a heat wave. It didn't take long to get back into the swing of things and I was soon in the routine of setting up the cloudbuster and recording in movie mode, taking full advantage of the warm summer days.

I would generally wait until the afternoon before starting, as by then the main body of the sun was behind the large yew trees that dominate one side of my research area. The foliage was so thick that you could barely see any light through them at all, but anything passing above the trees or higher in the sky would be illuminated by the sun's rays. I would set the camera up on the tripod and then aim and zoom in a little just above the tree tops, making sure I managed to get the top-most part of the tree in the picture. This would provide me with a visual reference point and plenty of space to catch anything that happened to be moving above the tree tops.

As I was mainly looking to record UFOs, I didn't take much notice of anything that showed up on the footage at lower altitude. Because the foliage and anything else that was normally green was now white, it could sometimes make it quite hard to see some of the objects that were shooting by, as many of them were also white as they reflected the infrared radiation. Looking back on it I was still very fixed on the idea of capturing UFOs similar to the ones that appeared in Trevor James Constable's books, so I tended not to take too notice of anything that looked small, or didn't look too much like a craft or some kind of biological UFO.

I would usually film in twenty-minute portions when using the movie mode—the main reason being the size of the files and the

Quest for the Invisibles

fact that when I transferred the footage on to my computer, the files were measured in gigabytes rather than megabytes. Doing it in twenty-minute portions made it much easier when reviewing several hours of footage. I would generally view the footage at quarter speed before slowing it down to one-eighth of the recorded speed or sometimes even slower, as it is surprising how easy it is to miss something when it is only captured for a fraction of a second. I have even gone back over old footage and found something I had totally missed when I first viewed it, possibly from being distracted or having tired eyes from reviewing of footage.

Ideally, I like to sit down alone at the end of each evening and go over the day's footage free from disturbances. I would spend many hours scanning and reviewing the footage in slow motion, and noting anything of interest that I happened to come across. I would usually take quite a few breaks while checking footage, as I would get eyestrain when working for too long. In the end, I decided to get my eyes tested and found that I needed to wear glasses for reading and computer work.

For some time I had been noticing fast flashes of light that seemed perfectly straight as they streaked from one side of the screen to the other, or straight up from the ground and shooting skywards at a colossal speed. I thought at first that they were insects or perhaps some weird reflection from the sun. Because I was fixed on the idea of capturing UFOs at the time, I tended not to take much notice of them, as I was seeking larger objects moving higher in the sky.

If anything of interest showed up in the footage, I would make a copy, leaving around ten seconds at each end of the footage. I would then keep the twenty-minute portion of the movie mode recording from which the footage came, and file it away for future reference, as well as taking stills from it if appropriate. All the other twenty-minute portions of infrared movie mode footage with nothing on were eventually deleted to free up space on my computer, as were the many hours of footage taken using my standard Exilim digital camera, which showed no signs of anything other than scores of insects and a normal view of the recorded scene.

Skyfish

By late August 2010 the weather was pretty dismal and there was a lot of heavy rain and grey cloudy skies daily. I found it impossible to film on such days, so took a couple of weeks off from the infrared work until the weather improved. Within a few days I was starting to feel a bit lost, as I was accustomed to the warm sunny weather and enjoyed the daily routine of trying to film invisible UFOs. I decided to sit down and go through all of the movie mode footage I had recorded so far, making doubly sure that I hadn't missed anything before going ahead and deleting the files from my computer.

I began going through each piece of footage taken since April 2010, which is when I first started to use the cloudbuster to help with the attraction process. This also was around the time I started noticing the fast flashes of light shooting across the screen. I began slowly trawling through all of the twenty-minute portions of footage in slow motion, sometimes going down to one-sixteenth of their original speed. It was a long process and very monotonous at times, but I was determined to account for these fast flying streaks that had been appearing over the last four and a half months. I knew they weren't insects as they were much too large, and besides, some of the speeds they appeared to be maintaining were like nothing I had ever seen before.

I found that insects barely appeared on any of my infrared footage, apart from the odd bee or daddy-longlegs, which sometimes came near enough to the lens to be recorded. I had the camera zoomed in and focusing on the top branches of the one-hundred-foot lime trees, so anything that flew between the camera and the top of the trees tended to be out of focus and wouldn't appear in the footage. Many insects were far too small and didn't seem to have enough body mass to reflect the infrared light, especially when they were several feet or more away from the camera. However, on the standard non-infrared camera that I often used in tandem with the Canon G10, I found that insects did show, but it was pretty clear that they were appearing at the end of the day.

I began to locate the flashes of light that had been appearing in my earlier footage one by one, and I made a note of the time they

Quest for the Invisibles

appeared so I could go over each piece of footage in slow motion and have a proper look. I was still unsure as to what they were, and was very keen to get to the bottom of the mystery. I continued going through the footage, carefully making sure I didn't miss anything. The process ended up taking much longer than I had anticipated as there were literally dozens and dozens of these fast moving streaks of light. In the end, I spent the best part of two days going through every bit of footage, and to my amazement I discovered that the culprits were actually fast moving skyfish, which is a term given to this suspected form of life that is not currently accepted by mainstream science.

I had already heard about the skyfish/rod phenomenon after watching a documentary about it on television. Although there has been lots of controversy and no physical proof since Jose Escamilla first announced the discovery, there was some convincing footage on the documentary I saw, and it was pretty obvious to me that they were real life forms. One of the bits of footage I remember seeing was of a large skyfish that looked like a missile or a torpedo as it flew through the sky, and I have since captured a few similar examples in my infrared footage. The ones I have filmed all had a dome-shaped front end, as well as a couple of long fin-like appendages at each end of the body, and there is nothing in their makeup that resembles anything that appears in nature.

I have heard many people try and explain these creatures away as insects, saying that they have been stretched out by the photographic process. This does happen, and I've seen footage taken of moths and other insects that did appear slightly stretched, and the resulting image did resemble a "rod." But from my experience this happens mainly at night or in a low light situation, which results in a slower shutter speed. I have also seen this occur on a few occasions when an insect has flown past the camera at speed and was close enough to be in focus. When the footage is carefully studied it is easy to determine that it was an insect, and not in fact a rod. I have found that when I am filming or taking photographs with the infrared camera, most insects get lost amongst the background of black and white. These colours are a

Skyfish

result of the tree's foliage reflecting infrared light, also combined with a large assortment of dark branches that are visible as well.

I must admit I was quite surprised to catch skyfish in the infrared, as up to this point I had assumed that they were physical creatures that were invisible because of the immense speed at which they fly. Apart from that, I hadn't given it much thought, as my real interest lay in trying to film UFOs. At first I felt bewildered at the sight of these strange looking creatures, and even a little disappointed upon discovering the truth (that they were not UFOs) behind the fast flashes of light I had been recording.

As time went on and I recorded more footage, the more I became intrigued by these strange looking life forms. They were mainly white in colour due the reflection of infrared radiation, and some of the first ones I managed to capture were very faint and almost transparent. In some of my earliest footage you could actually see the roof tiles through their bodies as they travelled from the apex of the roof and down towards the guttering, before disappearing altogether as they moved out of the sunlight.

Over the coming months I gradually developed an interest in them, and after looking in great detail at some of my live footage taken in the infrared, I began to notice certain similarities that appeared on the various different types. The footage I was taking on the infrared Canon G10 was filmed at thirty frames per second on warm days with lots of sunlight. From my observations I noticed that every now and then they appeared to arch their bodies, followed by a thrusting action combined with a flurry of wing power to project themselves forward at great speed. They were also able to perform twists and turns with hardly any effort at all, as well as changing direction at the drop of a hat.

I noticed that the fin sections of the skyfish, which often ran almost the whole length of the body, were sometimes the only parts that became visible to the lens of the infrared camera. In some bits of my early footage the body section of the creature could hardly be seen with any detail; in many cases the overall shape of the skyfish could only be determined when these fin sections were totally illuminated by sunlight.

Quest for the Invisibles

Although many of the first skyfish I recorded in the infrared were very opaque and only just visible as the sun hit them, they took on a whole new life when slowed down and studied carefully. Once the footage was slowed down you could actually see the individual parts of their bodies flexing as they moved and performed their high speed turns and manoeuvres. Featured in my Skyfish Serenade video on YouTube is a classic example of a skyfish as it makes a sharp ninety-degree turn skywards, at immense speed. The fins are not actually visible on this particular piece of live footage, as the sun was in the wrong place to illuminate the whole creature, and it looks just like a rod or a stick flying through the air at speed.

I have found that most insects that I have come across, such as midges and small flies, generally bunch together as they fly round and around, and tend not to move much from their chosen area. Skyfish, on the other hand, appear to travel perfectly straight and with purpose. In fact, I have never seen anything move that fast before. Bees and larger insects such as flying beetles also fly in a straight line at times, but they generally fly much slower and their constantly flapping wings easily get recorded on a camera or camcorder. Another thing I've noticed is that some of the skyfish shoot straight up vertically from the ground at speed, something I have never seen many insects do, as they mostly seem to rise gently up into the air at an angle, with their feet dangling underneath, similar to the path of a plane taking off.

The more I thought about it, the more I realised that it was just as Trevor James Constable had stated in *The Cosmic Pulse of Life* when he claimed that the existence of invisible life forms was essentially plasmatic, i.e., having their form expressed in heat substance. It seems like this applies to skyfish as well, as, after all, infrared radiation is heat. When using a standard digital camera the infrared is of course blocked by the hot mirror and all the invisible objects are absorbed into the normal sky background, so none of these creatures can be seen. I believe, though, that many of these modern standard camcorders are very sensitive—especially on a day where you are filming into a perfectly blue sky with little or no clouds, and they can easily record anything giving off a slight infrared or ultraviolet reflection. This is why digital cameras have been the first to reveal these creatures, on certain occasions.

I am convinced that skyfish are a reality, although in my opinion they are part of an invisible world that borders our own physical reality, which many refer to as the etheric. From my own observations and from reading various articles in the BSRA (Borderland Sciences Research Association) literature combined with various strange incidents that I have encountered over the years, I definitely believe there is an overlap of phenomenon from this etheric world. Most of it is invisible to the human eye, but every now and then someone or groups of witnesses will see something very strange drifting across the sky.

As with Kirlian photography which excites the life energy with high frequency energy in order to make it illuminate and reveal an aura, I believe many creatures and craft are sporadically being made visible to the naked eye by the constant bombardment of the skies with pulsed energy from radars and other sources, coming up from the planet Earth. This is a sentiment also shared by Trevor James Constable in his book *The Cosmic Pulse of Life*, and he has included a diagram showing Planet Earth – Pre Radar, as well as another showing Planet Earth – Post Radar, and the difference between the two is pretty obvious.

As the summer continued I spent as many days as I could outside filming, trying to capture as many different examples of skyfish as was possible, although by now it was becoming quite a laborious task looking back through three or four hours or so of footage in slow motion. It was really only the thrill and anticipation of what I might catch that spurred me to keep on and on. Ideally I would need a blue sky with a bright sun and preferably no clouds, unless they were small clouds that were way out of my line of view. These conditions were perfect for filming in the infrared, and the resulting footage was quite clear. In my early work I would often capture them as they flew from one side of my garden to the other, but later I began to use the cloudbuster to attract them into the immediate vicinity, where I could then film them more easily.

Quest for the Invisibles

I continued to experiment by pointing the cloudbuster just above the top of the three huge ewe trees that bordered one side of my research area; I found through trial and error that the best time to record was usually around mid-afternoon when the sun was at its strongest. I would aim the Canon G10 just above the tops of the tallest tree branches, and that way I was able to frame a large portion of sky in the shot as well. As the large skyfish flew over the treetops they were completely lit up just for a fraction of a second as they hit the full sunlight, and although some were still faint, others would have a more solid appearance. Some of them were five or six feet in length, and I came to this conclusion after comparing earlier footage taken of them passing near fallen branches, which I was able to measure. Other specimens appeared to be a foot or so in length, or sometimes much smaller.

Many of the larger skyfish I recorded above the treetops had appendages that resembled paddles, which they appeared to use to make those amazing high-speed turns. In many cases they had what resembled a fine gossamer-like fin structure that ran down either side of their main body, and this became clearly visible to the infrared camera as it caught the full sunlight at certain angles. These appendages appeared rather large in proportion to the body, and many seemed to have two at each end and two slightly smaller ones in the centre of their bodies.

Throughout the summer of 2010 I recorded quite a collection of different types of skyfish, all taken at thirty frames per second on my infrared converted Canon G10. After a while I had sufficient footage to put a small movie together which I called Hidden Reality 1: "Skyfish Serenade." I put some background music on the video entitled "An Ocean so Blue," which was written and performed by an Oxford band called The Red Bamboo. The song title fits well, as I often think of the sky as being like an ocean with us being the bottom dwellers.

Although I had been filming for only four months, I wanted to get the footage out there and show the world just how real this phenomenon is. There is some good footage on the movie, and it shows a wide variety of types. I have slowed the footage down in

Skyfish

many places, and this helps the viewer to see the different ways in which the skyfish use their fins to move through the atmosphere.

I considered 2010 to be an experimental year. So much of it was trial and error as I tried different attraction and filming techniques. It was a steep learning curve and I often filmed for hours, only to find that I hadn't captured a single thing in my footage. It was disheartening at times, but I knew that perseverance was the name of the game, and I wasn't prepared to give in.

It was amazing how many different types of skyfish I encountered, from the small, centipede-like examples up through to the much larger ones that appeared to look like bullets or missiles with domed ends. I filmed one of the missile types as it flew straight over the top of a large tree in my garden; it appeared silvery grey in colour and looked to be at least six foot in length or larger. The object was moving at such a high speed that it appeared on only one of the frames of the thirty frames per second available, and with its large, fin-like appendages on either side it did look very much like a missile to me. This and the bullet-type skyfish can be seen on the Skyfish Serenade video on my YouTube channel, as well as many other examples including a close-up of another type, showing what appears to be a segmented body.

I have found from many years of recording in the infrared that if you get the angle just right and are not too far away when filming, you can often get a more solid looking creature that sometimes appears light brown in colour. When filming from a distance with the sun blazing behind you, the result will normally be an almost pure white-looking skyfish, and this is due to reflected infrared radiation. At times it was almost as if some of them were made of glass, and they only became truly visible to the infrared when they were hit at a certain angle by the light from the sun. When filmed out of direct sunlight, many become more opaque and some of them hardly register at all on the footage.

Over the years I have amassed quite a collection of skyfish footage with my Canon G10, showing all sorts of different body and fin arrangements. After taking a good look at these strange life forms, it became obvious how well these creatures seem to fit into nature—just like all the creatures of our physical world. In my opinion, they are not using the air currents of our physical world to fly as our birds do, but seem to be moving through the ether. There appear to be quite a few distinct types, and I have put as many of these as I could into the Skyfish Serenade video.

Late summer was spent capturing more examples in the infrared, as I was totally intrigued by them and wanted to know more. As well as filming skyfish, I would spend about an hour at every session using the G10 to take photographs, as I was still trying to capture UFOs and other invisible life forms. Although it could be mundane checking hours of footage at slow speed, it was gratifying to come across a new type of skyfish or some other strange, invisible form that I had never seen before.

Eventually I thought about getting a camcorder to help me with my work in 2011, as this would then enable me to zoom in a lot further, and hopefully result in better footage. The zoom on the G10 was okay, and it worked well for recording just above the trees, but I wanted something that could zoom in towards objects that were at a much higher altitude. The plan was to find a good camcorder that wasn't too expensive, and then send it off to MaxMax in the USA for a full spectrum conversion. This would then give me the choice to view any part of the light spectrum I wished, by using different pass filters in combination with the camcorder.

After much deliberation I put the idea on hold in order to do some research and find out which models I could convert to full spectrum. I didn't want to buy a new camcorder only to find out later that I couldn't get it converted. So I put the idea in the back of my mind and kept using the Canon G10 to explore the infrared. Besides, I still had a lot more to learn about this strange infrared realm before I began looking into the ultraviolet. I had a good month or two before the sun would begin to fade and lose its strength, so I made the most of every sunny day and continued to record whenever I had the chance.

CHAPTER FIVE

Spinning Jennys and Plasmatic Gliders

During a bright afternoon in late August 2010, I was sitting in a chair looking at the LCD screen on the Handycam, when all of a sudden I saw what looked like a paper aeroplane flying way above the huge lime trees in one corner of my research area. I immediately looked up at the sky to see what it was, and there was no sign of anything. I thought it was strange, so made a rough note of the time when it appeared so I could later check through the footage in slow motion and try and locate it.

It was only ten minutes later when I saw three more similar objects, but this time they looked more like oblong pieces of paper although thicker, and folded in a v-shape. A couple of them disappeared momentarily, only to reappear a few feet further on. I had never seen anything like it before, so I stopped the recording and loaded the footage onto my computer for a closer look. I knew there was no way they could be bits of paper floating on the breeze, as it was a completely still day with no wind, and the lime trees were around a hundred feet tall.

I loaded the footage onto the computer, located the objects, and began watching them in slow motion. The first one looked much like a paper aeroplane, as I had first observed, but it appeared to be moving in a very controlled way, swooping up and down and slightly changing its course. Like the other three that appeared in the next piece of footage, it also seemed to disappear for a second or two—and it was quite astounding to watch. I found myself going over the same piece of footage again and again in total disbelief, and ended up just sitting there for a while trying to rationalise it and put it into some kind of context.

The three objects in the second piece of footage were all flying in a similar way, and they appeared to be v-shaped and resembled pieces of paper folded down the centre. They seemed to be using

Quest for the Invisibles

each side of the v-shape in a similar way to wings, adjusting their position slightly as they twisted and turned through the air. How they kept their momentum I do not know, but they were moving quite fast and appeared more than capable of gaining height when required. In another piece of footage that I obviously missed at the time, another example could be seen, and this was also in a v-shape but upside down compared with the other three. Again, it altered its course by changing the position of its "wings," and at one point soared up above the trees, before swooping back down and resuming a straight course above the treetops.

I was totally baffled by this footage and it didn't make any sense to me. They appeared to be life forms of some sort and moving with purpose across the sky, but I couldn't make out any other features apart from the paper-like body. I examined the footage more closely, looking for any sign of a mouth or eyes or anything else that stood out, but there was nothing more to them. During the following weeks I encountered a few more, and they eventually became a familiar part of my footage. Sometimes they would circle round and round the area where the cloudbuster was pointing and, as with the other invisible forms I had captured so far, there was no sign of them on the standard digital camera which I also used aside from the Canon G10.

As time went on, more and more of them showed up in my footage. I began calling them "plasmatic gliders," or just "gliders," mainly because of the way they would glide effortlessly through the atmosphere. It was an eye-opener discovering these strange creatures, and it made perfect sense that they must be a form of life existing in the realm of the infrared, alongside the many different types of invisible forms that I had already encountered.

Around this time I began to capture other life forms in my footage. Again they were brilliant white, but they looked like a mass of spinning energy. They appeared to fly in groups of usually half a dozen or more, and after slowing the footage down I could see that they were similar in shape to a stingray, with large triangular fins on either side of their bodies. Instead of using their fin-like structures in a flapping or undulating motion, as a flatfish would do, these strange creatures chose to fly nose down while

Spinning Jennys and Plasmatic Gliders

spinning the rest of their bodies around to gain momentum and lift, very much like a helicopter. It struck me as quite a labour-intensive way of moving, and it must have required a lot of energy.

I named these strange life forms "spinning jennys." The name suited them well, as the first noticeable thing you see when you encounter them is a mass of glowing, spinning energy. I went on to film dozens of them over the rest of the summer, and so far I have always filmed them in groups and have never encountered a single one flying alone. At first I thought they were white butterflies until I noticed they were quite large in comparison to the trees, and they were obviously spinning instead of flying. The first time I captured them they were moving in a small group, and were travelling at quite a pace just above the treetops. Along with the "plasmatic gliders," they appeared to be their own light source, and were just as bright out of direct sunlight as when they were in it. They are some of the strangest things I have ever encountered since I began looking into the infrared, and it never fails to intrigue me when I watch the footage of them spinning through the atmosphere.

After studying them for a while, I decided to make a replica out of some plasticene that I obtained from the local paper shop. I successfully matched it to some of the stills taken from the live footage and I even recreated the spinning motion, following every movement along with the images on my computer. As with the footage of the "gliders," I could see no sign of eyes or a mouth on the spinning jennys—or any other prominent, distinguishing features, despite having several close-ups of them.

The footage kept coming, and I decided to put together another video for YouTube entitled Hidden Reality: 2 "Gliders and Spinning Jennys," and this included some of the best footage I had at the time of these two very different types of invisible forms. The video features a couple of bits of footage showing large groups of these spinning oddities as they flew above the treetops in my garden, and includes one bit of footage where a lone "glider" can be seen flying alongside a whole group of "spinning jennys."

I became very interested in these fellow inhabitants of the invisible world, and couldn't help but wonder how many other

Quest for the Invisibles

strange things were flying around up there, undetected by mankind. This strengthened my resolve to capture as many different types as I could on camera.

I never filmed either of these life forms without using the cloudbuster as an attraction technique, as many of these creatures appear to come down from way up high and flock towards the area where the cloudbuster is pointing. I have often wondered if they are travelling in groups for safety reasons; just as in our physical world we have predators, so perhaps that is the case in the invisible, although I am yet to record such behaviour so it is only a theory.

One day I was in the garden aiming my infrared camera roughly in the direction of the cloudbuster, when a large "glider" flew straight towards me, and I had to step backwards to keep it in the frame. I've included this piece of footage on the YouTube video, and you can see in slow motion how it uses its "wings" to change course, revealing how they adjust their position. It was a very strange experience for me, as this "glider" was completely invisible except in the lens of my infrared G10. As I stood there looking into the LCD viewer with one eye, I could see nothing at all with the other. I was really excited knowing I had captured an example of one of these life forms almost head on, and at such a low altitude.

My sister stood with me during some of these encounters and saw absolutely nothing, despite the camera picking up dozens of them flying right over the area where we were standing. Besides using movie mode to record, I was also taking photos with the infrared camera, and would often walk away from my research area and shoot from different angles—in the general direction that the cloudbuster was pointing. I continued to catch more invisible objects such as fast moving plasmatic spheres, as well as cigar shaped objects and a multitude of other strange glowing shapes.

I carried on until the beginning of winter of 2010, at which point there was hardly any sunlight, so I decided to take a break until the warmer days of 2011 came around. During this time I decided to get a camcorder, as I had been thinking about it for some time now. I ended up getting a Sony DVD-650 Handycam, and later sent it off to MaxMax in the USA for a full spectrum conversion. It

Spinning Jennys and Plasmatic Gliders

didn't take long, and within a few weeks it had arrived back safely, along with the ultraviolet pass filter I had also ordered. I was excited, as I had always wanted a camcorder so after unpacking it I went outside and looked at the world in full spectrum for the first time.

To be quite honest, the full spectrum view of my garden didn't look a whole lot different from the normal view. The trees and the sky appeared lighter in colour, as well as taking on a slight luminosity, but apart from that I couldn't see much difference at all. Unlike the infrared camera that made the grass and foliage white, there was no sign of any of this on the full spectrum camcorder, and the normal range of visible colour seemed to dominate the overall image. Because the camcorder could now see the infrared and the ultraviolet as well as the visible spectrum, then in theory I knew it should be able to film invisible objects as well. I scanned the sky for a while to see if I could pick up anything interesting, but it was a dull day with large grey clouds, so the overall visibility wasn't that good.

I took the ultraviolet pass filter out of the small bag and held it up to the light. It was very dark and I could barely see anything through it. I then lit a match a few inches away from the filter to see how bright it would appear, and this time I could see the flame clearly, although it appeared to take on a reddish hue. I attached the filter to the camcorder using the screw fitting, which now meant that the camcorder could only see the invisible ultraviolet radiation, and the visible spectrum was now blocked. As with all ultraviolet pass filters, there is a slight infrared leakage or "bump," as it is often called, and this is due to the fact that the infrared is a harmonic of the ultraviolet. I have never considered this a problem, and always thought of it as an extra bonus. There is a filter available from MaxMax which you can put on the camcorder in combination with the ultraviolet filter, and this will block the infrared "bump," so you can then get a pure ultraviolet image if desired.

As 2010 gave way to 2011, the cold days of January and February set in and there was snow everywhere, along with the freezing cold nights. There was not much I could do at this point, so spent the time reviewing all the footage I had taken to see if I

Quest for the Invisibles

had missed anything. This was a long process, as all movie mode footage was recorded at thirty frames per second, and most of it was in twenty-minute sections. Many of the things I had already captured were moving at quite a speed, so they only appeared on a few frames, taking only a fraction of a second to move across the screen.

By this time I had amassed a huge amount of footage, and it all needed to be checked again, every single second of it. I generally delete my footage after checking it a few times, and this way it doesn't take up too much of my computers' memory. Twenty-minute portions are easy to check, compared with one or two hour segments which could get a bit monotonous. It took nearly a month to go through it all, but I eventually got the job done and managed to free up quite a bit of memory in the process, while finding a few things that I had missed.

As time went on I took more and more footage. The computer I had was slowing down and didn't seem up to the job, so I ended up getting a new one with a lot more memory. This did the job and made the memory issue a thing of the past.

It took a while to come to terms with the amazing life forms I had captured on camera, and it was frustrating that official science doesn't recognise life beyond this physical plane. As Trevor James Constable had mentioned in his book *The Cosmic Pulse of Life*, it is no good trying to get recognition from anywhere for this kind of work. Many people can't cope with the idea of creatures or strange craft flying around in the skies, even though many are invisible and appear to exist on another plane. The best I could do was to release some of the footage on my YouTube channel, which gives others the chance to take a look and make up their own minds about it.

Another good thing is that more people seem to getting out there and filming their own footage. I have met quite a few people via the Internet who are involved with work in a similar vein. A few years ago I became acquainted with a gentleman called Gregory Harold from the USA, and he managed to capture a whole load of interesting footage on a Kodak Super 8 camera. He had originally set the camera up in his garden as a security measure, whilst

Spinning Jennys and Plasmatic Gliders

trying to get to the bottom of some vandalism and other odd events that had been happening.

After reviewing the footage, he found the camera had captured images showing strange balls of light, as well as alien-like beings that appeared to be coming down into his garden via some kind of transporter beam, and lots of other strange phenomena. His subsequent book, *The Alien Connection,* as well as his DVD, *Harold's Mystery,* give a great insight into his story and are well worth investigating. We will hear more about Gregory Harold and some of his discoveries later on in chapter nine.

There are many people out there looking into the invisible, and although our methods may not be exactly the same, we all have the same aim in common, and that is to verify the existence of another dimension that overlaps our physical reality. The more people that get involved the better, in my opinion. As they say, "the camera doesn't lie," but many still find it hard to accept that there is another reality. After all, it is easier to say you don't believe in something if you haven't seen the evidence, but let us not forget that "absence of evidence does not necessarily point to evidence of absence."

The footage I present to you via the YouTube footage and in the photograph section of this book is one hundred percent genuine, and anyone can take such footage; it just takes time and patience, and a belief in your ability. I'm not out to deceive anybody; these things are as real as you and I, and yes, it is a revelation when capturing them for the first time on a camera or a camcorder. When it becomes a reality, seeing them day in and day out, the initial shock gives way to a gentle understanding, and your thought process starts to change, and many old beliefs are broken down. The reality of being in touch with another realm of existence with its strange phenomena will inevitably change the way you think, and once open to the new knowledge, it is hard to turn back ones' interest in the subject.

As mentioned earlier, a copy of Trevor James Constable's book, *The Cosmic Pulse of Life* is essential reading, and I wouldn't have reached the point I am at now without the knowledge gained from this and his earlier book, *They Live in the Sky*. I still read both

Quest for the Invisibles

books regularly, and there is always a copy of *They Live in the Sky* next to my bed, as I like nothing more that to read about Trevor's early adventures in the Mojave Desert. Judging by the price and limited availability of Trevor's books back in 2010, when I started trying to get my collection together, it seemed that there was sudden interest in his work. One bookseller told me they were like gold dust, and he had recently sold a hardback copy of *They Live in the Sky* for eighty pounds. I was just glad that *The Cosmic Pulse of Life* had been re-issued, meaning that thousands of people now had the chance to read about Trevor's work with invisible UFOs, which, in my opinion, is key to understanding the wider UFO enigma.

I carried on with my work in the infrared part of the spectrum, going out whenever there was a clear blue sky with hardly any clouds. Occasionally I would slip the X-Nite 330nm UV pass filter on to the full spectrum Handycam, but there was still not enough light coming from the sun to get clear footage. I knew the sun would be putting in more of an appearance over the coming few months and would be much higher in the sky, and this would be just what I needed, as the UV rays would then be that much more intense.

As I continued looking at other people's footage on YouTube, as well as getting to know a few more of them via the UFO scene, I realised just how many people like me were out there trying various methods to capture things on infrared cameras and camcorders. It makes me happy knowing there are lots more people out there investigating the invisible, and as more and more of us look beyond the physical realm, then more evidence will come to light, which is a good thing as far as I'm concerned.

You only have to watch some of the documentaries about UFOs that are on television and the Internet to realise that so many people are now filming strange objects in the skies using their camcorders—and even on their mobile phones, as is more and more common nowadays. Many of the new camcorders are so sensitive that they have the ability to capture strange looking craft as well as other objects, even without being converted; I have seen a whole host of interesting things over the years that people

Spinning Jennys and Plasmatic Gliders

have captured, particularly in places like Mexico, where some great footage has surfaced in recent times.

As the warmer days of early 2011 arrived, there was still not quite enough sunlight to use my full spectrum Handycam and UV pass filter, so I continued using the infrared Canon G10 to record and take photographs, knowing that it would be some time before I would be able to start exploring the ultraviolet. Although this was disappointing, I knew I had to be patient and bide my time until later in the summer, when there would be more ultraviolet light coming from the sun.

From March to May I continued working in the infrared and took some of my best photographs, one showing a tadpole-like "critter" flying through the atmosphere just above some nearby bushes, which is featured later on in the photo section. I also took two photographs of an aeroplane that was passing overhead, which also included a whole load of "gliders" at a lower altitude. I later captured a strange, feather-like invisible life form right above one of my neighbour's gardens, as well as a star-shaped object moving just above the ewe trees on one side of my research area.

Occasionally I found disc-shaped UFOs in my footage, although these were generally quite high up and appeared too small to include in the photograph section of the book, but despite this, the overall shape could easily be made out. This convinced me that these craft were clearly attracted by the cloudbuster, as they always seemed to be flying straight above my research area, as if checking out what I was doing. Although this was somewhat unnerving and raised all sorts of questions in my mind, it was, at the same time, exciting to know that they were up there and I was managing to capture them on live footage, as well as in some of my photographs.

During late May and early June I captured quite a few interesting things on the infrared camera, and in one of them there appeared to be a pyramid-shaped UFO that was similar in shape to the "ventla" type featured in *The Cosmic Pulse of Life*. It was quite high in the sky and only appeared in one photograph, with no sign of it in any of those that followed. I was also finding a plethora of other strange shapes showing up in my footage, and it was hard to

Quest for the Invisibles

tell exactly what they were; some of them were almost trumpet-shaped, while others were oval or spherical, as well as triangular.

Some days lots of these objects showed up, especially on hot days when the sun was high in the sky. It was obvious they weren't aircraft, as it was easy to identify planes with the naked eye since their wings could be clearly seen and most of them would leave a trail of some kind as they passed above my house. The other thing about aircraft is that they would always appear in multiple shots as they made their way across the sky, whereas most of the other objects seemed to appear in just one photograph. Sometimes I captured them alongside or near to passenger planes, but most of the time they seemed to be much higher up in the sky, with many appearing just on the edge of cloud formations or sometimes even in the clouds.

Summer was now here with the sun much brighter and higher in the sky, so I started experimenting with the full spectrum Handycam and ultraviolet pass filter. I spent a week or so using it in conjunction with the cloudbuster, putting it on a tripod and zooming in on aircraft as they passed over my research area. Although it was good practise, I wasn't catching much at the time, and I found that the resulting footage wasn't particularly clear and had a purplish tint to it. This affected the quality of the images, and not much detail could be made out in any of the resulting footage.

When I originally ordered the filter I hadn't taken any of this into account, assuming I could just screw it on to my Handycam and get clear ultraviolet images. As the days rolled by and I kept getting more and more blurred images, I wondered if it had been a mistake getting the filter in the first place. I felt disappointed, but hoped I would eventually work out the best way to use it, and maybe then I could finally begin my exploration of the ultraviolet realm.

CHAPTER SIX

A Glimpse Into the Ultraviolet

On June 11th 2011 I awoke to find the most glorious day with the sun high in the sky, and within an hour I was outside and ready to take advantage of the bright sunshine. Later that afternoon, whilst filming with the full spectrum camcorder in combination with an X-Nite 330nm UV pass filter, I filmed an object that looked very much like a jellyfish. I saw a flash of colour on the viewing screen, so I knew I had definitely caught something. The object was travelling tentacles first, as it moved at colossal speed just above the roof of my house, which I was using to block out the main body of the sun—basically, the sun obliteration technique. I was very excited and couldn't believe my eyes when I later loaded the afternoon's footage onto my computer. There it was, as clear as day—an atmospheric "jellyfish." It only showed up on two frames, and by the second frame it had moved far enough out of the sunlight that it was almost too faint to register.

This was just the boost I needed. Up to this point I had not been sure of the best way to use the ultraviolet filter, and had spent the last few weeks trying different things. I had tried using it in the same way as my infrared filter, aiming the Handycam into the area of sky above the trees, but due to the lack of sunlight, the images didn't come out well. I could now see how clear the images became when more sunlight was allowed to pass through the filter, so I began experimenting and using parts of my house roof to block out the main body of the sun, while still allowing a small proportion of the rays to just peek over the top of the roof. I gradually worked out the most effective way to film in the ultraviolet, making the best use of the ultraviolet pass filter to get much clearer footage.

The day I filmed the "jellyfish," I had zoomed in towards the edge of the roof that I was using to block the sun, and it seemed to be

Quest for the Invisibles

just in the right place at the right time and exactly in focus. With this success, it was now time to properly explore the ultraviolet. I knew from filming in the infrared that where there is one invisible creature there are normally more, so I now felt sure I was on the right track. I had also captured a few other invisible objects during that session, including a large, transparent, worm-like creature passing just above the roof of a nearby garage. As with most of the objects that I went on to capture in the ultraviolet spectrum, they were only visible to the lens of the Handycam the moment they were totally illuminated by the sun's rays.

The interesting thing was that I managed to capture these examples without using any method of attraction such as the cloudbuster, and it now became apparent that these life forms must be everywhere. It was a revelation, and as a result of this I began concentrating on the ultraviolet for much of the next three and a half years, whenever the conditions were right. I still continued to take photographs with my infrared Canon G10, and I was still managing to capture the odd, invisible form from time to time, some of which appear in the infrared photographic section of this book.

When using the sun obliteration technique to capture invisible UFOs, one of the things to take into account is the movement of the sun in relation to the wall or roof that you are using to block the sun with. I found that in late summer here in the UK the sun would rise quite rapidly, meaning I had to move the position of the camcorder slightly every four or five minutes, and this was to stop the full glare of the sun from overexposing and totally ruining the recording, and possibly damaging the camcorder's sensors. I always try to keep a small amount of sunlight towards the middle of the frame, and that way I can see the invisible object in its entirety as it passes above the roof.

I was very excited after capturing my first good piece of footage with the new 330nm UV pass filter, and it was a great feeling. I soon found that many of these creatures appeared to be moving just above rooftop level and sometimes even lower, and although I still continued filming in the infrared, the results I seemed to be getting in the ultraviolet were much more fulfilling and seemed to

A Glimpse into the Ultraviolet

show more detail than anything I had captured so far using the infrared-enabled Canon G10.

The skyfish that were showing up in the ultraviolet footage were very different in shape to the ones I had captured in the infrared part of the spectrum. Many were more fish-like in appearance. I couldn't help wondering if these invisible forms were moving through the actual air in their reality, or swimming through some kind of dense ocean in relation to their make-up, or perhaps flying through an ever-present, etheric reality devoid of such things. It raised many questions, which I thought about constantly, and I still think about it even now. In the end, I came to the conclusion that I should just keep filming and not worry too much about the whys and wherefores.

I continued on through the summer of 2011, and whenever I had a bit of spare time I would be outside filming. Most times I would capture at least something, although many great video recordings were ruined because they were slightly out of focus or the image was far too small in overall size. It takes a lot of patience to get good images, but over time I developed a feel for it.

I had nobody to guide me in my ultraviolet pursuit of UFOs. There were no experts, so I did the best I could at the time. I continually read Constable's books, especially *They live in the Sky*, as, like Trevor in his early adventures in the Mojave Desert in the late 1950s, I was following no rule book, and was having to depend on my instinct throughout the whole adventure. At times I felt I understood what Trevor must have gone through, especially when I was staring into the heavens with a cold breeze blowing in the early mornings I chose to go out and film. Deep inside I knew I had to carry on, refine my methods, and discover better ways of capturing footage.

My early morning adventures were the first time I had ventured out of my comfort zone for some time, being up and ready at four thirty in the morning for about a week. The days of camping and attending festivals back in the late 1980s were just a distant memory. As I got older I had developed a pattern of sleep that I rarely deviated from, but for those few mornings at least, I was up at first light and setting up my equipment.

Quest for the Invisibles

Once the sun had come up, I still had to wait a little longer until it was high enough in the sky and giving out sufficient ultraviolet light for me to get a decent picture. It was still chilly in the early hours, and as I had never been one for wearing gloves, I constantly suffered with cold hands. I would take a thermos of coffee out with me so I didn't have to keep running back indoors and wasting time, waiting for the kettle to boil. In the end, I did not have as much luck with my early morning ventures as I had hoped. Most of the time there was just not enough light coming from the sun at that time in the morning to use the ultraviolet filter.

I decided I would be better off waiting until later in the day when the sun was much higher in the sky. I eventually began filming at around midday, by which time I was able to block the sun behind the rooftops. I would set the Handycam up on a tripod and then sit down and keep a constant eye on the LCD viewing screen. Occasionally I would walk around taking pictures with my infrared camera, but this soon became hard work because I had to keep racing back and adjusting the camcorder every five minutes each time the sun rose above the top of the roof I was using. I eventually set the alarm on my mobile phone to warn me when the five minutes was up, so I could then reposition the Handycam.

As my ultraviolet footage was in five-minute segments, it was much easier to review than the twenty-minute infrared movie mode footage. Nevertheless, I would often record for three or four hours at each session, and this was still very time consuming. For the next few years I put all my time and effort into exploring the ultraviolet using the full spectrum Handycam and ultraviolet pass filter, as well as putting together a Quest for the Invisibles website, featuring photographs, information on filters and many other things. It worked out well. I had thousands of visitors to the site, as well as receiving lots of emails from interested people. Later on, I developed incompatibility issues and problems with certain search engines. Although it still looked fine on many old computers, the format was likely outdated so I decided to get rid of it and do a brand new site in the near future.

By the end of summer 2011 I was getting a better feel for the Handycam and was starting to capture some really interesting

A Glimpse into the Ultraviolet

footage. I seemed to be getting just one example of every different kind of biological UFO. It felt as though I was getting a helping hand from up above, as each one made its presence known to me. I continued on for as long as I could, into late October, but eventually the sun wasn't rising high enough in the sky for me to block it out with any of the nearby roofs or buildings, and the trees were also starting to get in the way. I could have walked around the local area setting up the Handycam on a tripod and using some of the other nearby buildings to block out the sun, or even some of the neighbour's roofs, but I felt uneasy about doing that—I was trying to stay pretty low key.

Over the summer months I recorded quite a few examples of skyfish, and a whole host of other biological life forms and strange objects. The skyfish I captured in the ultraviolet were much clearer than anything I had caught in the infrared, and this was due to them being at a much lower altitude. I found that I could film them just above the rooftop of my house or nearby buildings, and like the other invisible forms I was capturing, most of them were travelling from west to east.

Many of the new examples I recorded looked a lot more like the fish of our physical world, and many had clear fin and tail sections. They all appeared to have the same rod-like body which was barely visible at times compared with the fins and tails, but a few of them had what looked like a hollow tube as their main body. This was intriguing as it was something I had not seen in previous footage, and in one of the video stills, which I've included in the ultraviolet skyfish section, you can see this hollow tube viewed almost head-on—but one can only speculate as to its true purpose. This image has become one of my favourite skyfish stills, and I often find myself looking at it in total disbelief, even after all this time.

As the year moved on and the days grew colder, I packed my equipment away for the winter months. Although I spent most of my time in 2011 concentrating on the ultraviolet, I had still captured some footage in the infrared—but the ultraviolet footage was so much better. As soon as it started to warm up a little in early March of 2012 I began filming again, and although the sun wasn't yet at

Quest for the Invisibles

its brightest, I still managed to capture the odd thing here and there. I knew within the next month or two there would be a dramatic increase in both the sun's warmth as well as the amount of UV light coming from it, so it was only a matter of time until I would get better results using the Handycam.

Between April and September 2012 I went out at every spare chance, and just filmed and filmed. Sometimes I spent most of the day amassing a huge amount of ultraviolet footage, ready to analyse whenever I would get a spare moment. By late September 2012 I had managed to trawl through most of the footage, and found some very strange looking life forms and other weird objects, with most of them travelling at tremendous speeds.

As time went on I became quite good at judging the angles of illumination, as well as the amount of zoom to use to get far clearer footage. The problem one faces is that if you don't zoom in a lot towards the illumination zone, you may capture more frames overall, but the object may look very distant and the quality may suffer. On the other hand, if you zoom in too much the object may be blurred and out of focus.

I always focus on the part of the wall or rooftop that is blocking the sun, and then zoom in until I am just about in line with it. There is a fine line between getting the position right or not. I had to scrap hundreds of bits of footage over the years, as they just didn't come up to scratch. But that is the name of the game, and you need a whole lot of patience. In some cases I have filmed for two or three hours, and after checking the recordings on my computer have found I captured absolutely nothing. On other days I would go out for just fifteen minutes and end up with some great footage. I suppose it's just like fishing, you have good days and bad days and you never know what you are going to catch.

I began experimenting with the white balance setting on the Handycam, trying out different shades of green such as the garden fence and the faded green wood of the garden shed, trying to get a better overall colour to the pictures. I also tried using different parts of the lawn which is the most commonly used method, and it was amazing how each subtle difference in colour affected the overall

A Glimpse into the Ultraviolet

result of the footage. To clarify, this white balance experimentation was done with the ultraviolet filter attached to the Handycam.

The only downside to the Handycam was that if you used the white balance function, you couldn't then record using the sports setting, and I always thought it was a pity they couldn't be used at the same time. The sports setting helps stop the blurring which often occurs when filming fast moving objects, and this is what drew me to this camcorder originally, as it was advertised as a great model for capturing wildlife, as well as having an optical zoom capacity of x60, which was another good selling point.

When I first bought the Sony Handycam from a company in the U.S.A, they only had PAL models available. This was great for me living in the UK where we use PAL, but they offered no other versions such as NTSC for the American market, or any other format. A limited supply remained and the stock was diminishing fast, so I immediately purchased one after confirming that it was a model that MaxMax could do a full spectrum conversion on.

I continued experimenting with the outcome of different white balances, using a whole range of different shades of green, trying to get a good overall colour on the subsequent footage. Each setting varied and altered the colour slightly. For example, when I used the green waste bin to set the white balance, the footage on the Handycam had a pure blue sky and the objects filmed appeared almost see-through, with certain body parts surrounded with a white edge. I wasn't impressed with the detail, so began experimenting with the many other available pre-sets. In the end, the results were more dramatic using the automatic function, and it also brought out a great range of colours in the area surrounding the wall or roof where the rays of the sun are barely hidden from view. I didn't need to do much with the video stills apart from tone down some of the brightness and contrast.

The objects being captured in the ultraviolet seemed very different to the things I had captured in the infrared. They were also different in shape and seemed more solid by comparison. I was also getting good close up shots, and this kept me going for the next few years. All in all, I ended up with hundreds of reasonably good images taken from my ultraviolet footage, and the

Quest for the Invisibles

images I present to you in the ultraviolet photograph section of this book are what I consider to be amongst the best examples.

Now that I know how to capture the "invisibles" at close range using the sun obliteration effect, I should be able to get better and better footage as time rolls on. I have also made it my mission to get some good close-ups of the many craft that are flying through our atmosphere using both an infrared pass filter as well as my X-Nite UVR filter, in combination with the full spectrum Handycam. The UVR filter blocks the visible spectrum while passing its two invisible ends, i.e. the infrared and the ultraviolet spectrum, and it is very similar to the Wratten 18A filter which was favoured by Trevor James Constable in his later work.

Keeping the prices down was the goal from the beginning, with the aim of getting all the equipment as cheaply as possible. This proved harder than anticipated due to needing extra-long batteries for the Handycam and the G10, as well as requiring two more tripods, as mine were starting to fall to pieces. The overall cost of getting the camera converted to see infrared, as well as the Handycam conversion and the shipping charges, became quite expensive—but it was still well worth it. There are sites on the Internet showing how to do infrared and full spectrum conversions, but I have read many horror stories about people that ruined their cameras or camcorders with the do-it-yourself approach. As far as I am concerned, these things should be done in a clean room and by qualified people who know the process inside and out—that way you are guaranteed to get the job done right.

The sheer amount of footage I was able to capture in the ultraviolet was just mind blowing. I filmed many different examples of fish-like creatures over the years, and some appeared with what looked like just a mouth, with no visible side fins and hardly anything else except a tiny tail. I also recorded life forms that appeared to change shape as they came down from the sky, and one looked just like a smiling pair of lips on the first frame, yet as soon as it hit full sunlight it resembled a mushroom in its overall shape. By the next frame it had completely disappeared from view, well beyond the reach of the ultraviolet Handycam, which was just incredible. In the photograph section of this book you will be able

A Glimpse into the Ultraviolet

to see for yourself exactly what I'm trying to explain here, as sometimes a photograph can speak louder than a thousand words.

Another thing I became aware of through persistently filming in the ultraviolet is the presence of thousands of small particles that all appeared to be travelling in the same direction, as if caught up in some kind of flow. At the same time this didn't appear to affect any of the physical insects such as the small gnats that happily swarmed around the gable ends of the house on warm, still days. I found that most of the objects I was filming through the ultraviolet lens on the Handycam seemed to ride this invisible flow, which moves at quite a rapid rate.

I went on to create two more videos for YouTube, and they were entitled Hidden Reality 3 & 4: "The Ultraviolet Realm." These two videos featured much of the footage that I had discovered during my first few years of filming in the ultraviolet, and it went over quite well. On some of this live footage you will notice the hundreds of small bits of debris which suddenly become visible as they catch the full sun, and if you look carefully, it looks very much like an endless stream of energy that is constantly on the move. I decided to see what direction it was going in, and according to my compass it showed that the energy flow was moving in a west to east direction. I wondered if this was the etheric flow that I recall reading about, although I could not remember exactly where it was that I had first came across the idea.

I ended up flipping through *The Cosmic Pulse of Life*, frantically trying to find any references to the movement of the ether around the Earth. After about half an hour I found the chapter in which the etheric flow was talked about, and then I found the exact paragraph I remembered. In this chapter Trevor James Constable mentions the fact that Dr. Wilhelm Reich had independently determined the presence of this basic west to east movement of the ether, which he went on to name the orgone envelope of the earth. So maybe this was the flow of the ether after all, and these were indeed the etheric creatures that lived in this rather fast moving world.

It appeared as though many of the smaller creatures and objects used this flow to ride upon, although at times it seemed as if they

Quest for the Invisibles

were caught up in it and unable to break away as they were carried along, just as the smaller sea creatures of our physical world get caught up and swept along in whatever direction the current takes them. This is another reason that many of my stills taken from my live ultraviolet movie mode footage are showing these life forms side-on, because I am filming them as they race by on this flow from west to east. I don't recall ever catching anything head-on, although I have filmed quite a few invisible objects descending from the sky, as well as shooting up from the ground towards the sky.

The exciting thing about filming in the ultraviolet is that I never quite knew what was going to turn up next, and this is what kept me going. The huge amount of different things I was catching varied so much, as you will see in the photograph section later on in the book. Of course I realise I am only capturing a small amount of the many potential invisible forms that are out there with my limited work, and there is only so much one person can achieve whilst working alone. The sheer amount of footage that I obtained all had to be viewed at a much slower speed than it was recorded at, and this could be very time consuming. Despite this I have always found it enjoyable, especially when I happen to come across new images.

Sometimes it could mean spending about four hours just watching an hour's worth of footage at quarter speed or even slower, carefully looking for anything that appears out of the ordinary. As I have often said to the many who have emailed me since I set up the Quest for the Invisibles website, "blink and you can miss them," and that is so true. When you think how many frames your device is filming at—in my case the infrared Canon G10 films at thirty frames per second, and the Sony Handycam at twenty-five frames per second—and then you multiply that by the recording time, it soon adds up. The chance of missing something is always a possibility, especially when even a small glitch in the playback speed could result in overlooking something that is showing up for just a fraction of a second. One thing I learned early on was that you need to have a lot of patience doing this type

A Glimpse into the Ultraviolet

of photography, and it certainly isn't something that can be hurried in any way whatsoever.

Throughout 2013 and most of 2014 I didn't take many photos using the infrared Canon G10, as I concentrated more on the ultraviolet part of the spectrum, mainly because I was getting better results without the use any kind of an attraction method. There were still plenty of times when I filmed all day only to find out of focus images and nothing of any interest. It would take hours to go through all the footage only to find nothing, so it could be very disappointing at times. I realised that I was filming only a very small area of the sky, but then I would suddenly get some great footage, so continued using the same locations to record from.

The results were too good to ignore. I captured so many varieties of invisible life just above my garage roof, as well as above the house roof, that it was amazing. On one occasion, I captured three consecutive frames showing a rather obscure looking invisible life form, and I have included this in the photograph section. As I was filming it just above a nearby roof the object turned around, offering me a rare view of its underside. This revealed how its wing-like appendages were connected to some kind of a tube-like structure at the rear, something I had never seen before. This particular footage was taken while I was trying out different white balances, and that is why the sky appears bluish in colour, and the actual creature appears glassy white.

Most of the time I used the Handycam on the automatic setting instead of using the white balance option, as I found the footage that was recorded in automatic mode was much more pleasing to the eye, and it also brought a better quality to the overall picture. When using the white balance option, some features disappeared or were hardly visible at all, and in many cases they lacked the clarity of those taken using the automatic setting.

Because I filmed just above the rooftops of my house and nearby buildings when looking into the ultraviolet, I never managed to capture any types of craft. I found that the sun obliteration technique was great for illuminating things just above rooftop height or slightly higher, but anything much further up in the sky would become somewhat blurred due to the lack of sunlight. I tried filming to one side of the sun, hoping to bring more light into the scene, but this resulted in lots of reflections that would ruin the picture. I wasn't too disheartened as I still managed to get some great footage just above the house and nearby rooftops, and knew I could still learn more and record much more as it happened.

As mentioned earlier, more of my ultraviolet footage can be seen in the videos "Hidden Reality: Parts 3 & 4," which can be viewed on the QFTI YouTube channel. The videos contain lots of stills as well as live camcorder footage—all recorded during my first few years of exploring the ultraviolet realm.

CHAPTER SEVEN

Back To Normality

It had been a challenging few years since I began my quest to photograph invisible UFOs back in 2010. By now I had become a virtual recluse and my whole life had been totally taken over by photography, combined as well with the many long hours of sitting at a computer going over footage. My photographic quest certainly hadn't gone the way I had envisioned when I first started out, but my breakthrough into the ultraviolet during 2011, and the discovery of a whole range of different invisible creatures, had more than made up for that. I now found myself catapulted into a totally different direction to the one I was expecting, where I was now capturing lots of weird and wonderful biological UFOs rather than constructs of intelligent design.

I created file folders for each year since 2010 in an effort to keep everything neat and tidy; each one was full of infrared photographs and ultraviolet stills showing all sorts of strange, invisible phenomena. Although pleased with what I had accomplished, I was starting to feel burnt out by it all. I decided to take a break from UFO photography for a while and try and get back to some kind of normality. Up to this point, all I would think about was filming in the invisible. I often felt that I was losing my grip on reality—and at times even found myself questioning what reality actually was.

After a few months of taking it easy and catching up with the usual gardening jobs and other things that needed doing, I figured it was a good time to start the mammoth task of going through the large collection of notes I had taken over the last four years, and getting them into some kind of order. I had received lots of emails from all over the world from people with kind and encouraging words to say about my work, many of them repeatedly asking if I planned to write a book about my adventures in the invisible and if so, would I include photographs and explanations of the techniques used for infrared and ultraviolet UFO photography.

I couldn't believe so many people were writing to me. They all appeared to be genuinely interested in the subject of filming invisible UFOs. I had often thought that it might be a good idea to write a book. So many people were asking about it, that I felt compelled to sit down and write it. Some of the emails I received were very long, but I always took time to read them and reply. It felt good getting positive feedback about my work.

Every so often there would be a few weeks of dismal weather, often towards the end of the year, and sometimes I didn't get out and film for weeks at a time. Periods like this made me depressed, so I would sit down and read through the emails I had received over the previous year to cheer me up, and this inspired me to keep going. A large majority of them were from females, which surprised me. I didn't realise so many women were interested in UFOs.

It seemed that quite a few of these people had bought equipment after reading Trevor James Constable's book *The Cosmic Pulse of Life*, and many of them had later packed their equipment away after failing to film anything. After seeing my live footage on YouTube and the photographs that I had put on my Quest for the Invisibles website, they suddenly found a renewed interest and now the equipment was unpacked and they were asking me for advice. Many of them wrote to me saying that they felt inspired by my work, which was encouraging to hear, so I always did my best to answer their queries.

Writing a book now seemed like the obvious thing to do, but I knew it would take dedication and wasn't a thing to be taken lightly. I also knew that it wasn't going to be easy, but after much thought I knew it would be a good way to share my findings with others. I also realised that I would be providing more evidence to back up Trevor James Constable's pioneering work from the late 1950s, as well as the work done by the GRCU.

On May 2nd 2013, at the height of my filming in the ultraviolet, I was invited to appear on the Malcolm Boyden show on BBC Radio Oxford. Unknown to me, they were devoting the whole morning

Back to Normality

show to UFOs, as a load of UFO files had just been released by the UK government. As fate would have it, I had sent an email to the show on May 1st telling them what I was up to with Quest for the Invisibles, and had attached half a dozen photographs of invisible UFOs filmed in the ultraviolet. I didn't expect a reply, let alone an invitation asking me to appear on the next day's show.

As the day went on I forgot that I had emailed the show, and spent most of the afternoon cutting grass and tidying up the greenhouse, as well as planting my tomato plants into bigger pots. When I checked emails later that afternoon I was astounded to find an email from the editor of Malcolm Boyden's show, asking for confirmation that I could appear on the next day's program. She said they were looking for another guest who would come to the studio and talk about his or her experience of UFOs. I didn't need much convincing, and emailed back to say that I was more than happy to appear on the show, including my mobile number if needed to supply me with more information.

Within the hour I received a call from the show's editor and had a pleasant ten-minute chat with her about my photographic work in the invisible, as well as the newly released UFO files. I was scheduled to be on the next day's show just after the 11 am news, so my sister offered to take me in her car. On the way to the studio, which was a good forty-five minute journey, we had the radio on in the car and listened to the Malcolm Boyden show.

Every so often Malcolm would announce that Nik Hayes would soon be in the studio talking about the invisible sea creatures that he had managed to film in the skies over Radley and Abingdon, which did little to help my nerves. I had never referred to them as invisible sea creatures before, so I think Malcolm must have added that somewhere along the line. On reflection though, many of the photographs I emailed him were fish-like forms and one displayed the atmospheric "jellyfish" that I had managed to capture, so I guess that was a pretty fair description. I think he did me a great favour, as this helped to separate my photographic work from the scores of other UFO photographers who were focusing on capturing just craft.

Quest for the Invisibles

As we turned onto the A34 and got nearer to the BBC Radio Oxford studios, I remember feeling nervous and wondering what I was letting myself in for, and began questioning if I could really pull it off. Luckily we had left the house in plenty of time, planning to get there early enough so I could visit a coffee shop across the road from the studio and have a coffee and a bit to eat before I went in.

About half way to Oxford, I was listening to Malcolm Boyden having a chat with Professor Brian Cox, a rising star of television here in the UK. He talked about his reasons for naming his young son George Eagle Cox, in honour of the Lunar Lander, and also mentioned a project he was putting together that evening at the Sheldonian theatre in Oxford. And then all of a sudden, completely out of the blue, Malcolm asked Professor Cox what he thought about the fact that I would be on the show later, talking about how I had managed to capture invisible sea creatures in the skies above Radley and Abingdon. I waited intently to see how he would answer, knowing he would likely have to be careful with his reply.

Below is an excerpt from the radio show on BBC Radio Oxford on May 2nd 2013. I didn't put this part of the show up on the QFTI YouTube channel in case I got into trouble reproducing Professor Cox's voice without permission at that time, but my interview with Malcolm Boyden, however, can still be heard there.

Malcolm: *"And what do you think of my guest after eleven o'clock? His name is Nik and he is a UFO photographer, and he's discovered sea creatures that are invisible to the naked eye flying over Abingdon."*
Prof Cox: *"I would question him seriously over that."*
Malcolm: *"I intend to after eleven."*
Prof Cox: *"I yes, I um yes, what can I say?"*
Malcolm: *"Well, I'll question him."*
Prof Cox: *"Make him justify that statement."*
Prof Cox again: *"It would be one of the great scientific (pauses), it would be the greatest scientific discovery of all time. You're either talking to the person that changes the course of human history... or someone else. I'll leave it up to you to decide."*

Back to Normality

Malcolm: *"We'll find out after eleven. Isn't it exciting?"*

They both laugh and Professor Cox says *"yes,"* and he continues laughing along with Malcolm, sounding genuinely excited at the thought of invisible life in the skies. They then finish with a few niceties and Prof Cox tells of a new science series that is going to be on television here in the UK, and that was it—the interview was over between Malcolm Boyden and Professor Brian Cox.

Meanwhile, I was still sitting in my sister's car on my way to the studio. By now I was feeling even more nervous, especially after hearing what Professor Brian Cox had said to Malcolm Boyden about me. "It could have been worse," I kept telling myself. At least I wasn't dismissed as just another UFO crackpot. I know that my footage is one hundred per cent genuine, and I can say that with my hand on my heart, so I knew that ultimately I had nothing to worry about.

As we sat there in the coffee shop across from the BBC Studios I drank a couple of cappuccinos and soon felt a bit calmer about the situation. Ten minutes later we made our way over to the studio. We were given identification cards and my sister was ushered to an area just outside the main studio, I was then taken through a door and into Malcolm Boyden's studio. He immediately came over, gave me a strong handshake and told me it was an honour to meet me.

"Likewise," I replied, and told him I regularly listened to his show, and had been doing so for quite a few years.

"Sit down, sit down," he beckoned, as he pulled out a comfortable looking chair, then handed me a set of headphones which I put on just in time, as the familiar jingle they play after the news was just finishing. As a countdown started Malcolm glanced over and asked if I was ready, and that was it—we were live on the air and broadcasting to the whole of Oxfordshire and surrounding areas.

I must admit my heart was racing, despite the fact that I had done radio interviews when younger, as a singer in various bands. It was now 11 am and I wasn't feeling at my sharpest this morning, but it was too late to back out now—I had to keep cool and tell the

Quest for the Invisibles

listeners all about my quest to capture invisible UFOs using infrared and ultraviolet photography.

I can't recall the listening figures for BBC Radio Oxford around that time, but remember feeling sick with nerves when one of the studio guys told me earlier that thousands of people would be listening, especially today, as the subject was about UFOs. Up to that point I hadn't given much thought to the multitude of average listeners that regularly follow the show, and who would be hanging on my every word. My nerves soon subsided and I talked away, explaining as much as I could about the reasons for starting my quest, and describing some of the strange things I had captured on camera and camcorder over the last few years.

I talked with Malcolm for a good forty-five minutes, and he seemed genuinely interested in my work. He told the listeners several times that he had already seen some of my photographs that I had emailed to him the day before, and he agreed that they were pretty amazing. He later informed the listeners that later on in the morning some of my photographs would be appearing on the Radio Oxford Facebook page. He then invited the listeners to take a look for themselves and make up their own minds. After this we had a good chat. He asked some interesting questions, and it felt like I was talking to an old friend—that's how relaxed he made me feel. I felt as though I could have talked for hours with him and it was so easy to forget you were in a studio, and live on the air to thousands of people. All in all, it was a great day and well worth it; I had around six hundred extra visits to my website as a result of being on the show that day.

Malcolm also said he would love to have me back on the show in the future after I managed to get the book together, and this was something I was really looking forward to. Unfortunately, a few months later he quit the show and returned to football commentary. I know he is sadly missed by scores of people, including myself. He had a great way of making you feel happy and totally at home in the studio, and always did such good interviews and discussed numerous interesting topics on his show over the years.

Back to Normality

Much to my surprise, the entire interview with me was repeated on Monday, 6th May 2013, which happened to be a bank holiday. It was good to sit down and listen to the show from the comfort of my couch at home, even though it was strange hearing my voice coming over the radio. I was pleased with the way the interview had gone, and it had been great meeting Malcolm Boyden at long last. I had been listening to his voice on the radio for many years, but could now put a face to the voice. It would have been nice to meet Professor Brian Cox in person as well, so I could have had a proper talk with him. I've always enjoyed watching his programmes on television and have a few of his books as well.

I continued filming until late autumn 2013, and after the winter break carried on as usual until late autumn 2014. The sun was again putting in less of an appearance, so I began arranging all the notes I had taken, and started planning the book. The hardest part, which was recording the actual footage, was almost done. I had a good collection of footage for the photographic section, it was just a matter of sorting through it all and choosing the best examples. I still went out with the Handycam on sunny days, but I didn't seem to be capturing much.

Since getting hold of *UFO: La Realta Nascosta* by Luciano Boccone a few years back, I had been thinking about getting a Geiger counter for detection purposes. Most of the photographs in that book were taken from instrumental readings, and the GRCU used many different types of detectors ranging from ultraviolet, infrared, and temperature indicators, through to precision magnetic compasses as well as Geiger counters. From watching YouTube documentaries about Geiger counters, I eventually ordered one in early 2015 when they were advertised at half price on the Internet. I also ordered a few books on the subject to help me understand a bit more about radiation.

In Boccone's book there are some great photos that resulted from instrumental detection. Many were taken using a forty-second exposure, with some showing strange plasmatic life forms emerging from the midst of a glowing ball of energy. One of my favourite photos from the book is the UFO that appears at low altitude in Bangkok, Thailand. This was taken using colour infrared

Quest for the Invisibles

film. If it wasn't for the fact that the UFO was completely invisible, it could easily have been mistaken for a small airship. The GRCU certainly kept getting results, but they had around twenty-five members with lots of different cameras and equipment at their disposal. Some of them used infrared flash lamps in combination with their infrared cameras when filming at night. This proved to be very effective and resulted in much more detail in their photographs.

As soon as anything invisible was detected by their instruments, everyone would take multiple photos with their cameras, although they rarely saw anything until the photographs were developed. They filmed invisible, griffon-like birds and pterodactyl-like creatures flying high above the towns at night, as well as a whole host of other invisible life forms. Although they often took spontaneous photographs, they also experimented with long exposure times of up to five minutes or over, and this proved to be a very effective method. Some of the photographs that were taken by the GRCU are also featured in the photo section of Trevor James Constable's book, *The Cosmic Pulse of Life*.

The thought of filming at night was exciting after discovering the things the GRCU recorded, and a few years previous I had already invested in a CCTV camera with an infrared illumination beam. I had intended to use it for nighttime UFO photography, as well as general security purposes after a spate of break-ins in the local area. Due to time constraints I had not recorded much with it. I discovered through trial and error that if I turned the illumination beam down in intensity it made the overall image much darker, but orbs and other objects would still appear bright in the resulting footage. When there is too much light coming from the beam, then anything appearing in front of the lens tends to get watered down and lost in the swathe of infrared light, making it almost impossible to see.

I managed to film groups of orbs that looked about the size of small footballs moving across my garden one evening, with one of them rising slowly up in front of the screen as if it was playing up to the CCTV camera. Just after this, I walked in front of the camera and then moved off to one side. A few seconds after I was out of

Back to Normality

camera range a few orbs appeared, then more and more, until dozens of them were flying across the screen, which I found incredible. I saw absolutely nothing at the time, despite being only a few yards away from the camera when this happened. I have managed to catch a few other things on the CCTV camera as well, including a few amoeba-like life forms and other things such as long, luminous, rod-like creatures of various lengths.

I have always suspected that there are alien beings out there, and have always believed there has to be more than just us in the universe. When you look at the sheer size of the universe and see the number of stars and potential planets out there, it has never made sense to me that we should be the only planet with intelligent life. UFOs have been seen and talked about for centuries, and if you look carefully at some of the early cave paintings done by Neolithic man you will also see some that look just like spaceships, and a few even have objects that look suspiciously like ladders on the underside.

I have read many reports of pilots seeing massive UFOs which were, in many cases, picked up on radar and even the passengers saw them. There have been so many cases of fully trained and competent pilots with many years experience seeing flying discs and other unconventional craft, many of which were flying at extreme speed and manoeuvring like nothing they had seen before. I find it hard to believe these men didn't know what they were looking at, and I'm sure they at least suspected that these craft were not of human construction. In the early 1980s there were scores of reports of black triangular UFOs seen across the UK and Europe, as well as in the USA. All of this has been documented on television programmes over the years, and the reality of UFOs is considered common knowledge amongst the wider UFO community.

The foo fighters often reported flying alongside aircraft during World War Two by both English and German pilots, in my opinion, were probably spherical biological UFOs, similar to those Trevor James Constable filmed in the infrared, and appeared in his early book, *They Live in the Sky*. I have also managed to capture these highly versatile spheres on my infrared G10 during daylight hours, many of them travelling at high speed while others just flew slowly above the rooftops. I imagine that during World War Two they were disturbed by the sudden, increased activity of normally peaceful skies, with scores of aircraft and the noise of bombs and machineguns continually ruining their peace. In Trevor James Constable's later years, just before he gave up photographing UFOs in favour of weather engineering, he filmed many UFOs alongside the passenger airliners he was travelling on, so it seems we are entering their domain when we cruise above the clouds on the way to our destinations.

When I was in my early twenties I used to venture out with one of my mates at the time. Every Wednesday evening we had the night off from our girlfriends, so would drive out to remote places on the high ground near the wireless station and huge wireless masts near Daventry. We were convinced that something was up there, and would spend hours scanning the sky for anything that looked like a flying saucer. We did this for many years, but don't think we ever saw anything apart from the odd, strange moving light, but we took it seriously at the time.

Even back then, I believed in UFOs and was certain there was something out there, despite never seeing anything. It probably came from all the science fiction programmes that I grew up with such as Dr. Who, Star Trek, Lost in Space, and all the other space-orientated programmes that were on television at the time. That is why the Christmas day evening UFO I encountered while walking my dog was such a milestone in my life, as it proved beyond all doubt that they were a real phenomenon.

The night sky is also something that has fascinated me since I was young, and I used to spend many hours standing in my garden just staring up at the stars, totally amazed by the sheer number that could be seen on a clear night. One of my mates lent me a telescope back in my school days, which I later lost somewhere along the way, but even so, I got hours of fun out of it by zooming in and bringing the stars nearer to me. Even back then I always wondered if there was life elsewhere in the universe, and this inquisitiveness has followed me into adult life and been a major factor in my exploration of the invisible.

Back to Normality

In the next chapter we will take a look at some of the methods I used when filming the "invisibles" with both a camera and camcorder.

CHAPTER EIGHT

Techniques for Filming Invisible UFOs

There are a huge range of life forms and other strange objects hiding in the invisible, both at high altitude and also at a much lower level. It is probably best that we are oblivious to them all as they fly overhead, or else taking a walk up to the shops could be a rather unnerving experience. Even though I didn't doubt that there were various craft up there in the skies above planet Earth, it wasn't until I became familiar with the work of Trevor James Constable that I learned there were also invisible biological UFOs sharing the atmosphere with us. At the time, this was a revelation. But now, a few years on, I have a whole collection of my own photographs showing such objects—all of them taken using somewhat standard, available equipment.

Once one makes the decision to film invisible UFOs, you have a choice of using an infrared converted camera, or maybe a full spectrum camcorder on which you can then place external infrared or ultraviolet filters. It is all down to personal choice, and of deciding what will be the best method for you. MaxMax have a whole list of camcorders that they can convert to full spectrum, as well as cameras. As mentioned earlier, they also have a great range of different pass filters available in many different sizes, covering both the infrared and the ultraviolet, as well as the joint infrared and ultraviolet UVR pass filter.

Ultimately, it depends on what you intend to capture using your camera or camcorder. If you want to film "critters" or UFO craft without an attraction method then you will have to practice "seeing" the invisible, by using techniques where things are looked past rather than focused on. The idea is to defocus ones' eyes while looking up at the blue sky rather than staring at it, and this will enable the visual eye ray to interact with and detect any invisible objects. It will take a bit of time and practise, but will prove to be an

Techniques for filming Invisible UFOs

effective way of picking up invisible forms in the atmosphere. In many cases I try to pick up on any areas of the sky that appear slightly distorted or rippled, similar to a heat haze effect, or barely visible white streaks that you may be only just able to make out.

If you think back to the earlier section about Trevor James Constable and his early work in the Mojave Desert of California, you will remember how he objectified and photographed his first invisible UFO—the "amoeba" photograph. He could "see" a shimmering variation to the otherwise perfect sky right above, and then proceeded to take multiple shots of the general area, despite nothing being tangible in the regular ocular range at any time throughout the whole episode. This can be hard work, especially here in the UK where the sky is often grey or half covered in clouds, so when I first started my quest to film invisible UFOs I would normally get as far away from built-up areas as I possibly could, in order to get the widest possible view of the sky.

Remember, however, that many of these invisible UFOs travel at extremely high speed, so you have to learn to react quickly with your camera or camcorder. Another thing to remember is that these things tend be way up in the sky and many cameras do not have the needed zoom capability, so a camcorder can often be the best tool to use. It is still worth taking lots of photographs of the horizon, or certain areas of the sky, just to see if you can catch anything. This is something I refer to as the "lucky shot technique," and sometimes you can get results this way. It is also worth taking a few snaps of aircraft as they pass overhead, as many UFOs have often been seen and photographed near to them. In one of my photographs featured in the photo section, I took several pictures of a helicopter as it passed over my research area, and in one of the shots a strange invisible form can be seen just above it.

I sometimes get hit with a sudden overwhelming urge to start taking a series of infrared photographs, whilst scanning the area of sky immediately above. One instance when this worked well was on September 8th 2010, a few days before bad weather started setting in. On this day I had been using the cloudbuster and filming for most of the afternoon without having much luck, so packed the equipment away after deciding to call it a day. I sat down on one of

Quest for the Invisibles

the garden benches and poured a cup of coffee from my thermos flask when I felt an overwhelming urge to take a few more snaps of the area directly above me—despite the fact I couldn't see anything visually up in the sky. I aimed the camera into the mass of streaky clouds above the house and took four photos, one after another, before going inside to load the day's footage onto my computer.

On the computer I confirmed that I didn't catch much during the afternoon session apart from a few misty looking objects, which were very faint and far in the distance. I then came to the last four photographs taken, and to my amazement three of the photographs contained an image that looked like a craft of some sort, moving slowly, at low altitude, across the sky directly above me. Each shot was taken at $1/15^{th}$ sec, and I didn't use any zoom at all when I took these photographs in quick succession. I saw no indication of its presence, not even a faint flash of light, but despite this I had somehow managed to detect the object as it passed above me and capture it on the infrared camera.

In the first photo I got the impression that the object was moving diagonally from the top left to the bottom right of the screen, but in the next two photographs the wing shapes had all but disappeared, leaving just one small bright strip visible to the infrared. This remaining strip, which I thought was part of the object's wing structure, was actually moving towards the top right corner of the screen—in the total opposite direction to my first assumption.

I wrongly took it for granted that the large oblong features were wings and the long, thin missile-looking part was the main body, so I assumed it was moving as an aeroplane would in an obvious direction. Two of these three photographs are featured in the infrared photo section later in the book. Although I have captured a few good photographs this way over the years, I would find that I had reacted far too slowly with the camera many times, and found nothing on the subsequent photographs. The fact is that many of these invisible "critters" and craft travel at a tremendous speed and height, and by the time you click the shutter, the object is usually

Techniques for filming Invisible UFOs

long gone, so you have to be able to react quickly the moment you detect something.

There are lots of people all around the world doing what I am doing and that is great—the more the merrier. I was asked by an author a while back if she could mention my work in her new book, as well as on a few talk shows she was appearing on in the USA, including Coast to Coast. I didn't have any problem with that as it would introduce my work to a few more people, and it was pleasing to know that someone was interested in what I was doing.

She informed me that she had already appeared on a few shows in the past, talking about the pioneering work done by Trevor James Constable, and she wanted to continue on with the theme of invisible life. I am a big fan of Coast to Coast and I listen when I get the chance as it covers some great subjects. Because I live here in the UK and can't listen to the show as it goes out live on air, I have to wait a while for the shows to be archived and put on YouTube before I can access them.

Many people have written to me saying they don't have much luck in the infrared part of the spectrum, and this is something I also encountered at the beginning. I would scan the sky, half expecting to come across something that would be flying in my vicinity, but seldom managed to record any footage this way. I also tried taking multiple photographs of the horizon through a three hundred and sixty degree range several times, hoping to capture something. This is ok when using a digital camera, but it could be very expensive using a standard camera with infrared film, especially if you then have to pay to get it developed rather than develop it yourself. When I first began reading Trevor's books and learning how he managed to film UFOs in the Mojave Desert, he made it all sound so easy when he described the technique for seeing the invisible, but in the end it took me a good few years to gradually learn how to do it effectively. Even now I continue to practise the technique, and although I have had success on a few occasions, there are still times when I can't seem to locate anything for want of trying.

Quest for the Invisibles

The Sky Fishing Technique

One of the first techniques I used in the early days of my photographic endeavours in the infrared, after the "lucky shot technique" failed to provide me with sufficiently good footage, was what is commonly known as the sky fishing technique. This involves setting up your camera or camcorder on a tripod and leaving it there to film whatever happens to come your way. Once you have set your infrared converted camera to movie mode or fitted your full spectrum camcorder with the appropriate filter, it is just a matter of aiming into the sky and then waiting for something to fly into your location. The most important thing to remember is to keep an eye on the position of the sun. This way, you will avoid any unwanted reflections, which can spoil potential footage.

Even if you are using a standard camcorder that hasn't been converted to full spectrum, you should still be able to capture skyfish and other invisible forms, especially on bright sunny days when there is a blue sky with lots of infrared and ultraviolet light around. I proved this to myself when I first began using the Sony Handycam prior to its conversion, and went on to capture half a dozen flying disc-like objects way up in the sky, as well as a few other strange objects—neither of which showed up on the standard camera I was using in movie mode at the same time.

When using the sky fishing technique I have always found it best to film above a line of trees or just above some buildings, as this will help to block out the main body of the sun, help illuminate anything flying overhead, as well as providing a physical reference point in the footage. For best results I have always filmed on days when there is a totally cloudless sky, or at least use areas free of any clouds, as this will provide more clarity to the overall recording.

It is always best to spend a little time experimenting with the position of your camera or camcorder, making sure that you are not too far from the trees or buildings you are using to block the sun with. Once happy with your position you can then zoom in a little to bring the top of the tree line or roof a bit nearer. The key is not to zoom in too far, that way you should be able to get a good view and record several frames of any object passing by. Although

Techniques for filming Invisible UFOs

this can be an effective method if you have lots of time on your hands, it can be a very slow and time-consuming way of getting footage. You will have to keep an eye on the sun's movement in relation to the treetops if it is still rising, and then adjust your camera or camcorder accordingly.

The problem with filming here in the UK is the lack of flat land with good long views in either direction, as well as the less than perfect weather. This is a factor that will be different for everyone, and depends on your geographical location. With the area around my house, the horizon is blocked by buildings and trees, which means I am usually filming straight up into the sky above the trees. I have often wondered how many things I am missing because they happen to be flying below the tree level, or perhaps at head height or lower. Although I have had some success using the sky fishing technique, I eventually realised that it wasn't the most effective way to capture repeatedly good footage.

Attraction Methods

I started getting more results in the infrared only after I began using my version of the Reich cloudbuster as an attraction method, and this proved to be a much better way to capture UFOs and skyfish. Lots of things seemed to be attracted to my vicinity and began showing up in the subsequent footage, and both of my infrared YouTube videos were filmed this way. As we have already learned, cloudbusters are more than capable of dissipating clouds as well as attracting UFOs of all shapes and sizes. Dr. Reich also became aware of their presence when he started experimenting with a cloudbuster many years before Trevor began his work. The sky is such a vast expanse, and when you are only looking at a very small part of it, you have to really make your presence known in order to attract any potential UFOs to your position.

As mentioned earlier, Trevor James Constable started using the Star Exercise as a means of making his presence known when he originally set out to capture invisible UFOs at his Mojave Desert site, along with Dr. Woods. This proved to be a good technique, but I believe it is more effective if there are a few people together.

Quest for the Invisibles

Some can scan the sky and try and "see" anything invisible, as well as take photographs, while someone does the Star Exercise. I didn't find it that easy to do the exercise and get ready with the camera at the same time, and could have used a helper.

Trevor later began using a cloudbuster instead of the Star Exercise as an attraction method, and found it just as effective. His photos of invisible UFOs prove this beyond doubt. There is still quite a bit of controversy over the use of cloudbusters, and even Constable once felt driven to finance his own report on the matter. It seems that the weather is easier to manipulate with a cloudbuster than many people thought, and he was concerned over their rising popularity. The effect of multiple cloudbusters being used by lots of people who didn't fully understand the implications was a growing concern to him. He felt their use could not only affect weather patterns, but severely alter them.

I have seen some very crude designs for cloudbusters on the Internet, some of them consisting of just a few pipes and a bucket. The pipes were placed inside the bucket and held in place with resin, and the bucket was then filled with water. Although this is a very simple design, I do believe if quite a few of these were left at various places, they would have some kind of negative effect, with the probability of creating floods in normally dry areas. I have always been mindful when using my cloudbuster for attraction purposes, as I am only too aware of the consequences, and this is why I have not gone into too much detail about their construction.

I no longer use a cloudbuster today, as I later discovered that so much could be recorded without the need of an attraction process. I tend to use just the camcorder fitted with an ultraviolet pass filter, which, combined with the sun obliteration technique, works really well. You can also use an infrared filter to film like this, but many of the things that I have caught in the infrared using this method are almost see-through. Many are hardly visible unless they happen to be their own light source. I found that the objects I could capture in the ultraviolet are much clearer, and also appear more solid.

Another attraction method which I discovered much later, was using the infrared illumination beam of a CCTV camera. Although this can only be used in nighttime situations, it is still a very potent

Techniques for filming Invisible UFOs

device for attracting invisible forms. The model I ended up getting was the IRCam-1000, which has a fifty-metre infrared illumination beam, and was perfect for my needs. I bought the whole thing as a package, which included a DVR recorder and a decent sized monitor, along with the CCTV camera and the various leads that were required.

I have now secured my CCTV camera to a length of guttering, which allows me to stick it quite far out from my bedroom window, providing a clear shot way above the treetops that surround that part of my garden. I fix the end of the guttering into one of the drawers I have below the window, then secure it all down with some ballast so it points at about forty-five degrees. I can see the infrared camera view through the monitor next to it, and can zoom in manually should I see anything interesting. I aim just above the treetops, making sure the topmost branches are in the shot to provide a physical reference point in the footage.

As mentioned earlier, the GRCU group from Italy, led by the late Luciano Boccone, filmed pterodactyl and griffon-like creatures flying over the town of Arenzano in Italy in their infrared nighttime footage, as well as many other types of strange phenomena, so we know that these things are out there. When you think of it, there aren't too many infrared light sources aiming up into the sky, as most aim towards a gate or an entrance. I'm hoping that in the future I might catch more interesting footage using this method, although my preference is for daylight filming. I am still interested in exploring the nighttime inhabitants of the invisible world and when I get the time, would like to concentrate my efforts on this.

Detection Methods

Apart from using the technique for "seeing" the invisible, there are also a few other detection methods that can be used in conjunction with your photographic equipment. I recently started experimenting with a Geiger counter in daytime situations, and I have already filmed a few things with it—although it took a while to get used to holding the camera with one hand and the Geiger counter in the other. The RD1706 Radex Geiger counter I use

Quest for the Invisibles

measures in micro Sieverts per hour, and I have set the detection alarm to go off at 0.20 micro Sieverts per hour, which is an effective setting.

The GRCU also used infrared flash lamps in conjunction with their Geiger counters when filming at night. This would light the area up for the photographers to blaze away with their cameras. I may get one of these in the future for my nighttime work, and they are available from MaxMax in the USA. In many circumstances the infrared light from the flash lamp can extend your view into the infrared by up to 100 metres. This will enable you to catch more with your infrared camera, as many objects filmed at night are invisible to the infrared unless they are illuminated by infrared light.

If you are filming at night, the camera should be on a tripod, as this will avoid any slight movement that could affect the quality of the photograph due to the slower exposure time. If you are using the camera in movie mode or happen to be filming with a camcorder, this isn't as important. Most camcorders and cameras have an anti-shake setting, which helps keep the footage more steady.

The Italian group GRCU also had a whole host of detectors which often went off simultaneously when an "invisible" was present, or when a number of them were passing near their research area. These included ultraviolet detectors, precision magnetic compasses, and instantaneous temperature indicators, as well as frequency, light, ultrasound and ultraviolet detectors, all of which can be bought over the Internet easily. As I am working alone, I do not want too many different instruments for fear of being bogged down with too much technology. The GRCU, on the other hand, had many members so it was easier for them to control a wide variety of detection devices simultaneously.

The invisible world is full of strange creatures and objects that have probably been since the dawn of time, and they are all out there for you to photograph and film. Remember though, the more footage you record, the longer it will take to go through it all. This is another reason I began filming in smaller segments of half an hour or less. It is always a good idea to take a break every half hour or

Techniques for filming Invisible UFOs

so to give your eyes a rest, as it is surprising how easy it is to miss things when looking through tired eyes.

If you intend to use the sun in any type of obliteration technique, remember that the sun always moves so you must keep an eye on it and be ready to reposition the camcorder slightly every now and then to avoid getting too much sunlight in the scene. Remember to always use the LCD viewer on the camera or camcorder when working, and never look directly at the sun through the eyepiece. Although filters may appear dark and unable to let light through, don't be tempted to look at the sun through them—they are designed to pass ultraviolet and infrared, and this can end up causing you permanent eye damage.

The Sun Obliteration Technique

Back in mid-2011 when I first began experimenting with my Sony Handycam combined with the ultraviolet pass filter, I didn't know how to use it effectively. As mentioned earlier, I had tried using it in much the same way as I did the infrared camera by aiming and scanning certain areas of the sky, but soon found that when I filmed out of the sun the subsequent images were a bit blurred and not very clear. It was a very dark filter and I began to doubt its effectiveness. After repeatedly finding out of focus images, I wondered if I would ever be able to film anything with it.

The answer appeared to me one afternoon in early June, 2011, when I captured the atmospheric "jellyfish" as it moved above my house. Just by chance the Handycam was slightly zoomed in and dead in line with the top of the house roof and, unknown to me at the time, was in exactly the right position to capture anything passing just above. The sun had moved behind the house and was illuminating the topmost part of the roof, when all of a sudden I a saw a bright flash of light moving across the viewing screen from right to left. At first I wasn't sure what it was, although I had a feeling that it was an invisible form of some sort. At the time I didn't know if the image was going to be in focus or not, but thankfully, after loading the footage on to my computer, soon found that it was.

Quest for the Invisibles

This was how I discovered the best way to use the ultraviolet pass filter, and began to realise that I could capture footage without the need for an attraction method. I found that at a certain angle everything—including bits of dust and even masses of tiny insects—were illuminated by the sun's rays. This is what I now term the "illumination zone." Anything invisible entering above or through this point becomes illuminated and its form is revealed for a split second to the lens of the UV Handycam. I zoom in so am almost in line with the roof or wall that is blocking out the sun. It is not always easy with a regular camera unless you position it much nearer, so for this reason it is better to use a full spectrum camcorder.

The things I captured came in a whole range of forms; some had wings while others had fin-like structures like fish, and some appeared as a mass of swirling energy without any real form and bore no resemblance to anything I had ever seen before. I never knew what was going to turn up next, which was exciting—and that is why I spent about three years investigating the ultraviolet realm.

Ultimately, the amount of time I devoted to filming was dictated by the weather here in the UK. The best time to film was when the sun was at its hottest and quite high in the sky, and here in the UK that is from early June through late September. As a reminder, you must keep an eye on the viewing screen every five minutes or so, to ensure that the sun doesn't rise too high in the sky and ruin the picture. It is okay to let a small amount of sunlight rise above the roof you are using, and you may notice in many of my ultraviolet shots, the sun can be seen lighting up the edge of the roof line.

With time, the filming process became easier and the results better. The hardest part was trying to judge how far to zoom in to get a clearer picture. It's all about experimentation, and finding out what works best for you. Once you have located the correct angle to illuminate the "invisibles" you are set to go, and it is just a matter of time before something will pass by. Anybody can capture such footage and at the end of the day all it takes is patience, determination, and the will to get out there and do it. The most important thing is not to give up. If you fail to get any results at first,

Techniques for filming Invisible UFOs

just keep going and try different techniques and approaches until you do.

Capturing "Invisibles" on a Standard Digital or Film Camera

It is possible for one to capture invisible UFOs at night using a standard digital or film camera, in combination with a detection device such as a Geiger counter. I often take both the Geiger counter and my Casio Exilim digital camera with me when I go out for my evening dog walks. As soon as the Geiger counter alerts me to any variation in radioactivity, I start taking photos in the immediate vicinity.

I have found that manifestations pertaining to an invisible realm are objectified in most cases when the Geiger counter begins reading and exceeding radioactivity rates of 0.20 micro Sieverts per hour, even if these peaks last for only a fraction of a second. It is not always possible to react quickly enough with the camera, and in some cases I fail to record anything of interest; but for some unknown reason the flash is able to "light up" many invisible forms, allowing them to register and thus be recorded.

When using this method I set the camera to "burst" mode rather than just taking single photographs, and this can often show the progression of any invisible forms as they move through the atmosphere, which will allow you to get a much more detailed understanding of their movements. It will also enable you to have a view of the scene before and after anything appears, and that way you will know that you have not recorded mere reflections from static objects, such as stationary cars or reflective gate posts that happen to be in the vicinity.

Using this method I have captured many examples of coloured orbs, as well as a few other strange light manifestations. On a few occasions I have also recorded phantom mists that have suddenly formed in front of me and appeared to cover the whole road. Sometimes these mists can be seen dispersing in each subsequent photograph, although they can appear in just one shot.

An image captured on a film camera will often look different from one taken on a digital camera. This is because the image that

Quest for the Invisibles

appears on the photograph is dependent on how the film's emulsions react to the particular type of energy of any invisible form present. The same applies with infrared film, and Trevor James Constable found that many of the images he captured on photographs appeared as a result of the UFO's energy *nullifying* the film's emulsions, rather than *reacting with them* in the normal way.

I have also come across people using EM detectors such as a K2 meter to detect variations in electromagnetic fields caused by invisible phenomena. Many paranormal investigators use these when ghost hunting, and from what I have seen and heard they are quite efficient at detecting invisible forms of different types. There are also quite a few variations of EM detectors available over the Internet, and it is just a matter of studying them and deciding which one would work best for you.

CHAPTER NINE

Conclusions

To claim I have all the answers and that I have arrived at final conclusions about the invisible world around us would be wrong of me. I realise that I've barely scratched the surface of this vast subject. If I could perhaps one day get a full spectrum, high-speed camera and use ultraviolet and infrared pass filters with it, or maybe even get to work alongside one of the many University departments here in Oxford where I am based, then I am sure I could gather and present some incredible footage to the general public. Being able to shoot with more frames per second would show much more detail, and would enable me to see more of the body and wing movements of many of these invisible life forms as they twist and turn through the sky.

I intend to carry on filming and trying different approaches. I have been doing this for too long to walk away from it—it has now become my life's work. The footage I have recorded in both the infrared and the ultraviolet merely backs up the work done by Trevor James Constable in the late 1950s, and by the Italian research group GRCU in the late 1970s, and it makes me very happy to feel I have contributed to the cause. What makes it even better is the fact that our footage is all very different, meaning that a whole lot more has been revealed, instead of the same things being presented time and time again.

As for the many insect theorists out there who spend their lives trying to disprove that rods/skyfish exist, I think my footage clearly shows that they are just one of many types of invisible life forms that exist in our skies, alongside all the other examples I have managed to record. None of them look like any type of known insect species that I have ever seen or come across in any book that I have ever read, or am ever likely to read, I imagine. As mentioned earlier, all of my footage was taken during daytime hours on warm sunny days and was recorded at 30 fps on my Canon G10, and at 25 fps on the full spectrum camcorder.

Quest for the Invisibles

None of the insects I managed to record on rare occasions in the infrared and the ultraviolet were elongated in any way, as they were far enough away from the lens, and it was clear to see that they were insects. This stretching effect is more likely to happen in footage taken at night or in low light situations, but it can also happen during the daytime when insects pass by at speed in close vicinity to the camera lens.

Footage taken on a camcorder that uses interlaced video recording can stretch and elongate images of insects and even cars as they zoom by, and I have even seen it happen when filming a green tennis ball that was thrown at speed and recorded on my camcorder. On the subsequent image the tennis ball was noticeably elongated, and it was pretty obvious what had happened. I have seen the same effect happen to a football that was kicked at speed into the air, and on the resulting footage a slight elongation could be seen on the image.

This is not to say that all of the images presented as skyfish are actually skyfish, and I suspect that some of the footage that has been put forward is actually footage of insects that have been elongated by the recording process. I recently conducted a series of experiments where I purposely recorded flies and mosquitoes flying through the air from around a few feet away, using my Casio Exilim digital camera in the movie mode setting. In the subsequent images I found that the insects did appear elongated by quite a noticeable amount, and the multiple repeats of the wings on either side of the elongated body did give the appearance of a rod-like creature with fin-like appendages on either side. I have included some of this insect footage in the photographic section so readers can compare them with the images of skyfish.

The images that are featured in the ultraviolet skyfish section of this book, especially figs 37-41, show nothing that resembles multiple repeats of wings attached at regular points to an elongated body. On the first three images presented in figs 37-39, what you see is a fish-like body with fin and tail sections, and these are clearly formed and do not appear to be the result of an elongated insect body with repeated wing shapes on either side of the body. Fig. 40 also seems to have its own unique shape, with

Conclusions

what looks like two curved shark fins on either side, making up its tail section, as well as another curved shark-like fin which appears in the centre of its rod-like body. In all my years of coming into contact with flying insects, I have never seen any that resemble this life form, or have features such as wings or appendages that look anything like this.

The image presented in Fig. 41 also has a very unique body shape, with its flowing curves and an almost star-shaped tail section. I've captured a few images of this type of invisible form in the ultraviolet, and they look identical in every way to this image. Again, this life form does not appear to be made up of an elongated body with repeated wing shapes.

In Fig. 42 we see an image of a skyfish shooting straight up from behind my house roof towards the sky. Even in this image, one will notice that the only fin sections that are identical are the ones on either side at the rear of the skyfish. All the other sections are different shapes and sizes, and if the fin sections were merely multiple repeats of insect wings, one would see equally repeated images on either side of the body at regular intervals, especially as an insect flying in this position would be flapping its wings in tandem at a considerable rate in order to gain the height and momentum needed to take it skywards.

Also seen is an invisible void where the body section should be, as only the fin sections are reflecting enough ultraviolet light to make them visible to the lens of the Handycam. I have never seen this invisible void appear with such clarity or with such sharp straight edges on any insect images I have ever recorded in the ultraviolet, and I have always found that the whole body and wing sections are usually equally illuminated by the sun's rays as they pass by. Another thing to bear in mind is that once these life forms leave the illumination zone, they become so faint that many disappear, something I don't ever recall seeing the insects of our physical world do.

When you capture a daytime skyfish in the infrared using a camcorder, you get one straight image completely balanced on both sides by its appendages and various fin-like structures. When multiple images are taken of the same creature you can see the

Quest for the Invisibles

whole fin structure working together and changing in overall shape as the skyfish turns or accelerates. Examples of this can be seen in the Hidden Reality YouTube video, "Skyfish Serenade," which appears on the Quest for the Invisibles YouTube channel.

The images of skyfish I present in the ultraviolet section of this book are, in my opinion, part of the invisible world. This is backed up by the fact that they didn't appear on the standard digital camera I was also using in movie mode in conjunction with the Handycam. Another thing I have found is that to get clear footage like this of skyfish in the ultraviolet, I had to use the sun obliteration technique to illuminate them, and this involved getting exactly the right angle to film them from. The skyfish, along with the many invisible objects that I have recorded, are also capable of moving many times faster than any insect I have ever seen or managed to film. As I've said many times, they very often fade out and become barely visible to the lens of the camera or camcorder the moment they cease to be illuminated by the sun's rays. This is a feat that is unmanageable by their physical counterparts that always remain visible and appear constantly solid, and are often much darker in colour by comparison.

I've also found that the gnats and other small flies that are often found near the gable ends of my house appear to be totally unaffected by the west to east etheric energy flow that becomes visible in the ultraviolet camcorder footage. Most of the invisible forms that I have managed to capture however, all seem to be riding on, or in many cases appear to be caught up in, this fast moving flow. I have never seen any evidence of this energy flow whilst filming in the infrared, even when I have been filming exactly the same scene in both infrared and ultraviolet at exactly the same time.

The thing I've found most intriguing about the life forms I have captured in the ultraviolet and in the infrared is how perfectly adapted they appear to be for their environment—just like the creatures of our own physical world with their different wing and fin arrangements. Many of them bear no resemblance at all to any type of creature I have come across or read about, not even in the many science fiction programmes that I've seen over the years. It

Conclusions

has always amazed me how nature appears to fill its vacuums with life, creating creatures that fit perfectly into different environments, and as far as I'm concerned, the life forms that appear in both the infrared and the ultraviolet are no different.

I recently saw a good example of this during a nature documentary about marine life, which mentioned how they had discovered aquatic creatures existing only inches away from boiling water flowing up from hydrothermal vents at the bottom of the deep ocean. It just goes to show that life will exist and often flourish in seemingly inhospitable areas, and before this was filmed many would have thought it impossible that life could exist in such a hostile situation.

Although many of the "invisibles" appear to fly through our skies, an equal number that I have filmed use fins and tail movements to propel themselves (rather than flying), just like the fish of our oceans. While many of these strange creatures take these forms, there are others that appear as glowing energy or fast moving spheres of light, with no apparent means of gaining momentum. Despite this, they seem perfectly capable of moving at many different speeds, as well as making intelligent decisions as to the direction in which they move through the atmosphere.

As I mentioned earlier, I have only scratched the surface of the invisible world, and the many examples I have featured in the photo section of this book represent a small cross section of the many life forms that must exist in the infrared and ultraviolet. The more I continue with this work the more I believe that perhaps our planet and even other planets in the solar system are home to many different levels of existence, and not just the physical reality that we exist in. It has often crossed my mind that there could be multiple layers of life in many different dimensions, with all of them existing at the same time. I am convinced that the things Trevor James Constable, the GRCU and I have captured are examples of life from these other dimensions.

On top of all this you have the physical manifestations seen by many people, and in some cases filmed on their own equipment. I have seen countless documentaries on television about UFOs, as well as the many programmes that are now appearing on the

Quest for the Invisibles

paranormal and ghost hunting. This awareness seems to be growing, along with people's desire to document these strange happenings on their own cameras and camcorders, a feat not possible until recent times. The digital age has certainly helped many people get the answers they seek.

I have been contacted by quite a few paranormal groups over the last few years, but I tend to keep away from areas I don't specialise in, like ghost hunting and others. I have grown a bit wary after experiencing some strange occurrences—and I don't want to start opening doors that I can't close, which I am aware can happen sometimes. I am kept busy enough just trying to film UFOs, without having to deal with other types of footage. Trevor James Constable has mentioned that when he first started on his quest to photograph UFOs after meeting George Van Tassel, he began channelling various entities, some of whom were piloting the UFOs he was trying to film, when out in the Mojave Desert.

This was all well and good until he found himself in a battle to remain in control of his own body. He described how one evening he was propelled down the street like some kind of mechanical man, totally at the mercy of a force that had taken over his limbs. Luckily, after some help from his good friend and accomplished occultist, Franklin Thomas, Trevor was able to take back full control of his body. This just goes to show, you don't always know what you are dealing with when you are trying to attract UFOs and communicating telepathically with the being or beings on board. I have always been cautious and ask for protection and guidance before I do any of my photographic work in the invisible, as I don't know who or what I am inviting into my presence.

Out of all the different methods available to film these invisible life forms, I have found the most effective way to capture them is by using the sun obliteration technique. This will help to illuminate them so as to get clearer footage. Obviously this method works well on objects that are just above the trees or above the rooftops, but attraction techniques are needed with the larger bio forms and craft that are much higher in the sky. Detection tools such as a Geiger counter can also give you an upper hand in the invisibility stakes, and they can alert you to their presence.

Conclusions

On a few instances I've been out walking the dogs late at night when all of a sudden they would both stop dead in their tracks and immediately the Geiger counter would start going off. It is obvious that the dogs can sense something that I can't see, and they seem to be a good indicator of certain types of invisible manifestations. I have found though, that in order to actually record invisible objects at night, the area really needs to be lit up with infrared light, and I have never caught anything using just my infrared-enabled camera.

As mentioned previously, the flash from a digital or film camera is sometimes able to make many of these invisible objects register enough to be recorded, such as orbs and a few other strange light formations. It appears that the certain type of energy that orbs are composed of is strong enough to make the Geiger counter peak at 0.20 micro Sieverts per hour or more.

Although these peaks only last for a fraction of a second, if I am quick enough with the camera I can often capture a clear image of them. Very often the orbs are illuminated by the flash on the camera and appear in a whole range of different colours and patterns, but I have found that when I take photographs without the flash or use the movie mode setting to record, they don't seem to show up on the subsequent footage.

I have often wondered if the many invisible objects I have captured appear solid to each other, as we do to each other in this physical world, or if they can see or even sense us in our physical realm. Maybe they are totally oblivious to our presence and unaware of our physical world. I really don't know the answer to that question. I have noticed that many of them appear to fly over or around the trees and the house roofs, and this makes me wonder if they are interacting with these physical objects and avoiding them, as I have never recorded anything actually flying through solid objects.

I have had one experience, however, where a couple of bright spinning orbs about the size of a small football actually manifested in my house one evening, and then proceeded to go through one of the walls and totally disappear—only to reappear a few seconds later in my kitchen. That was the first time I had ever encountered

Quest for the Invisibles

anything like this, and it was immediately after I had been out with my infrared camera and had been asking for any invisible forms to make their presence known to me.

I found the whole experience very unnerving, and it just goes to show that you have to be careful what you ask for, as you might just get it. Over the years I have often asked for "invisibles" to appear in my photographs, and have sometimes been successful (my wish was answered). After that experience, however, I had to question whether it was a good thing to do. I didn't get a good feeling during the experience and ended up worried, as I was completely on my own at the time.

That there are also craft-like objects up in the skies seems almost a thing to be taken for granted nowadays, as media is forever pushing the idea of UFOs and aliens in its advertising, and most people I have talked to believe beyond doubt that there are UFOs up in the atmosphere. Documentaries on television have featured pilots of both military and commercial aircraft that have reported seeing strange craft flying near to, or even alongside their planes, and in some cases the passengers onboard had also witnessed the event. The Internet is full of footage captured by people using camcorders or cameras, some of it quite convincing, and if just one percent of these recorded sightings were true then this shows there is definitely something going on.

Some of the UFOs seen and recorded by people are not all craft, and a good example of this appeared on a BBC news report one evening a few years ago. The report announced that the Mexican Air force had released infrared footage taken by pilots of an aircraft, showing up to eleven invisible spherical UFOs that were being picked up on their radar, but could not be seen by the pilots. The incident happened when the plane was carrying out a routine anti-drug trafficking surveillance flight over the coastal region of Campeche. One of the pilots turned on the infrared camera after first noticing the objects showing up on radar. The UFOs ran rings around the plane—one minute they were behind and then the next minute alongside, before finally ending up in front of the aircraft. The fact that the Mexican Air force released this information is

Conclusions

amazing in itself, when most other establishments appear to be trying their best to keep it all secret.

I have captured a few craft-like objects over the years on my infrared camera, and some appear in the photograph section of this book. Unlike aeroplanes or helicopters, these UFOs are completely silent and don't register on any audio recordings. The shape of planes and helicopters are easy to notice, and they can be clearly heard and identified as aircraft, despite being so high in the sky. I often take infrared photographs of them as they pass over my research area, and as mentioned earlier, you can see an example of a helicopter with an invisible UFO just above it featured in the infrared photograph section.

One day back in August 2011 at around midday, something caught my eye. From my position on the ground it looked like a brown ball flying at around a thousand feet up, and I wasn't sure if it was the same object I had noticed several times over the previous few hours. It appeared to be travelling on a constant circular course, but never passed directly above me on any occasion. It appeared to be keeping its distance. As luck would have it, I had just unscrewed the Handycam from the tripod after removing the ultraviolet pass filter and was ready to pack it away—so immediately started filming and recorded the entire event in full spectrum.

After getting the object in my sights, I zoomed in until I had a nice clear view. The video of the object I encountered that day is on the QFTI YouTube channel for all to see, and is entitled *Strange Craft*. It basically appeared ovoid in shape, with two extensions that came out from the main body and looked much like radar dishes. Just as I focused in on it, the object slowly turned around as if it knew I was filming, and this I found quite unnerving. I tripped over the lawn edging and nearly fell over. I lost it temporarily, but managed to keep upright and continued to record. I got the object back in the sights of the Handycam, but from this new angle it looked totally different—appearing almost teardrop shaped.

This was very strange as I had assumed the object was ovoid in shape and would appear the same from any angle, just like a hot air balloon, which always looks the same no matter which side you view it from. In the *Strange Craft* video you will be able to see what I am talking about, and the difference in shape can be easily seen in the footage. I ended up making a model of the object out of plasticene so that I could better understand what I had captured. That wasn't as easy as it sounds, and I struggled to get the shape exactly as it appeared in the video.

The object seemed to be under intelligent control and was able to turn around and accelerate away at amazing speed, which it appeared to do when I started filming it. I ran as fast as I could to the other side of my research area and towards the edge of the garden, which had a good view of the sky towards Oxford. I was way too slow and just managed to glimpse the object as it disappeared into the clouds and was gone. It wasn't until I played the footage back to a few friends that I began to understand the importance of what I had captured.

Anyone looking up into the sky would have seen a dark object with no real features visible, and I doubt they would have given it a second thought. One has to wonder how many more of these things are flying around undetected in the sky. Before I became interested in UFOs I hardly ever looked up at the sky whilst walking, and have noticed that most people don't study the sky, apart from the occasional glance. Even when someone is driving a vehicle they notice things in the distance, but not immediately above—so in theory, many objects travelling across the sky could well go unnoticed by most people.

I don't know to this day what this object was, and what it was doing flying near my research area. I wondered if it was some kind of unmanned probe keeping an eye on my cloudbuster operations, or perhaps a small craft of some type. It was hard to tell, and suppose I will never know. I have shown the camcorder stills and live footage taken that day to so many people, including several connected with the military, and they are just as baffled as me when it comes to explaining it.

If anything can be concluded from my footage in the infrared and the ultraviolet, it must surely be that there are creatures, or things that resemble creatures, moving through the atmosphere, as well

Conclusions

as a host of other strange forms, including craft-like objects. I will keep looking into these invisible realms and putting the results on the QFTI You Tube Channel, as I like to share the footage and let people make up their own minds about what they are looking at.

When I began my quest to film invisible UFOs after reading the books by Trevor James Constable, it seemed so easy in theory, but as the inhabitants of the invisible world slowly made themselves known to me, it became very real and sometimes a bit disturbing. I found it hard to come to terms with the things I was recording, as many appeared to be alive. From reading Constable's books, I already knew there were biological UFOs out there, but I wasn't ready for the sheer numbers I began to capture in the infrared and, later on, in the ultraviolet.

Even NASA managed to capture dozens of invisible UFOs with their specially designed ultraviolet camera, and this happened aboard the Colombia spacecraft back in February, 1996. This event has become known as the "Tether Incident," and occurred during Columbia mission STS-75, which attempted to deploy a Tethered Satellite System. The intention of the tether was to absorb power from outer space, which would increase the satellite's endurance and cut down on operating costs. Unfortunately for NASA, the 12-mile-long electrodynamic tether snapped and after unfurling, became a 12-mile-long white, energized cord which attracted dozens of UFOs. The UFOs were very similar to the amoeba-like "critters" that were photographed by Constable back in the late 1950s, when he was filming in the infrared part of the spectrum from his site in the Mojave Desert.

The UFOs, however, were not visible to the astronauts in the Columbia, despite some being at least two nautical miles in diameter. You can see the footage for yourself on a videotape entitled, "EVIDENCE: The Case for NASA UFOs," which is produced by Terra entertainment, and narrated by David Serada. There are also segments from this video available on YouTube in case you have trouble tracking down a copy of the videotape. I have seen the footage many times, which is interesting and well worth viewing.

Quest for the Invisibles

Something else that gives credibility to the existence of invisible life is the discovery made during 2015 by the Thunder Energy Corporation, a breakthrough technology company featuring three cutting edge technologies in optics, nuclear physics, and fuel combustion. Thunder Energies, led by Dr. Ruggero Santilli and Dr. George Gaines, announced that it had recently detected invisible entities in our terrestrial environment, using the revolutionary Santilli telescope. The telescope was originally created for the detection of antimatter galaxies deep in space, and the Thunder Energies has previously confirmed their apparent existence, along with antimatter asteroids and antimatter cosmic rays.

On September 5, 2015 at 9:30 pm, Ruggero Santilli aimed a pair of 100mm Galileo and Santilli telescopes into the night sky above Tampa, Florida, with the intention of searching for antimatter galaxies. Because of sudden cloud formation, Santilli had to abandon his search for antimatter galaxies, and instead proceeded to aim the telescopes horizontally over Tampa Bay.

During this change in orientation, and also to his great surprise, Santilli noticed images of unidentified yet clearly visible entities appearing on the screen of the digital camera attached to the Santilli telescope. Although these entities could be seen without any enlargement, they were not visible to the naked eye and did not appear on the screen of the digital camera attached to the Galileo telescope.

After this unexpected discovery Santilli began using the pair of Galileo and Santilli telescopes in a dedicated search for more of these entities, which he called *Invisible Terrestrial Entities*, or ITE. It was found that with the Galileo telescope, normal matter-light is focused in the camera, yet, antimatter light is dispersed by its convex lens into the walls of the telescope. With the Santilli telescope, however, ordinary matter-light is dispersed by its concave lens, but antimatter light is focused in the camera, thus permitting images to be recorded.

Following systematic tests, Santilli detected the existence of at least two different types of ITE, known as ITE-1 and ITE-2, and he acknowledged that there could well be additional types that may be identified in the future. The entities that are classified as ITE-1

Conclusions

are also known as *dark* ITE due to the fact that they leave dark images in the background of digital cameras attached to Santilli telescopes, and the entities classified as ITE-2 are known as *bright* ITE because they leave bright images in the background of digital cameras attached to Santilli telescopes.

Although these entities appear to be emitting antimatter light, it is predicted that they are in fact made up of ordinary matter. This is because it has been established in particle physics laboratories that matter and antimatter particles "annihilate" at mutual contact, so this implies that that the existence of antimatter entities within our atmosphere would ultimately result in a cataclysmic explosion, due to matter-antimatter annihilation.

It is hypothesized by Santilli that the emission of antimatter light from ITE-1 could well be evidence that they are achieving locomotion via the acquisition of antimatter in their interior, with consequential use of anti-matter propulsion. He also hypothesizes that they achieve invisibility via the emission of antimatter light as a kind of exhaust. In his opinion ITE-2, by contrast, achieve their invisibility to the human eye as well as to conventional telescopes via engineering means capable of inverting the index of refraction of ordinary light.

Santilli also adds that both types of these entities have been recorded moving in the sky over sensitive civilian, industrial, and military installations, as if they were carrying out unauthorized surveillance. Although it is not clear what these entities are, Santilli states that being a scientist, his duty is only that of reporting the documentation of the existence of ITE-1 and ITE-2. He feels ultimately that their identification or lack thereof, belongs to the US government.

Although these are not entities in the true sense of what most people would expect, the fact that they do appear to be able to move at will through the atmosphere and position themselves above sensitive areas, is strongly suggestive of intelligence and free will. The fact that Santilli states that the emission of antimatter light by ITE-1 could be due to some kind of exhaust from antimatter propulsion, might well point to them as being living entities that happen to have this kind of propulsion as part of their actual make-up.

This is very interesting to me, as many of the invisible objects that I have filmed and photographed appear to move at various speeds, without any clear way of achieving this movement. They seem more than capable of changing both speed and direction at will, which shows intelligence and the ability to rationalise their surroundings and make judgements based on this. Could it be that they too are using some type of antimatter propulsion in order to move through the atmosphere?

Also provided in Santilli's paper are several photographs showing images of both ITE-1 and ITE-2, which were recorded in the evening sky over Tampa Bay, Florida, on digital cameras attached to 100mm and 150mm Santilli telescopes. In a few of the photographs taken, using a 15-second exposure, evidence of the slow movement and rotation of these entities can be seen from photo to photo.

As well as the photographs taken by Santilli there are also three other photographs presented that were taken independently by K. Brinkman in the night sky above St. Petersburg, in Florida. These show images of ITE-1 recorded using a digital camera attached to a 150mm Santilli telescope, and evidence of the movement and rotation of the entity can also be seen on these three images— taken using a series of three rapid shots.

There are other photos taken by Santilli that are featured in the paper, showing images of ITE-2 complete with enlargements of the original photographs. In one photo, a rather bright and unusual looking ITE-2 appears to be in the process of releasing seven smaller, but equally bright, ITE into the atmosphere. One of the most striking features of this photograph is the sharpness and clarity of the image, which was taken using a 15-second exposure.

More information can be obtained from the scientific paper of R.M. Santilli, "Apparent Detection via New Telescopes with Concave Lenses of Otherwise Invisible Terrestrial Entities (ITE)," American Journal of Modern Physics, **http://www.thunder-energies.com/docs/ITE-paper-12-15-15.pdf**. You can also obtain

Conclusions

it from the scientific archives of the R.M Santilli Foundation's web site at: **http.//www.santilli.foundation.org/news.html**.

Interview clips about this discovery can also be seen on the Business Television site at **http//www.b-tv.cpm/thunder-energies-discovers-invisible-entities/**

This is exciting news and provides more evidence that reality consists not only of things we can see and observe, but also of invisible things that we can't see. The scientific paper is well worth reading and provides very interesting information on this groundbreaking discovery by Dr. Santilli and the Thunder Energies Corporation.

More evidence of an invisible dimension surely comes from the discoveries made by Gregory Harold from the USA, the gentlemen I became acquainted with via the Internet, and whose work I briefly mentioned earlier, in chapter five. Greg's story starts after he and his wife moved from Michigan to Florida back in 1968. After moving into a new house that was built for them, a bizarre series of events began to unfold over the coming years, starting with some of their shrubs mysteriously dying overnight after being covered in an unidentifiable white, foamy substance.

Over time, Greg noticed several acts of minor vandalism which began to escalate. He found damage to his car including puncture holes in the tires, and a strange foamy liquid that appeared to have been thrown all over the car trunk and chrome bumpers. Unable to get to the bottom of the damage, he got a security camera that used super 8mm film. After mounting the camera on a bracket overlooking his back yard, he ran it on certain nights, hoping to catch on film the people or persons responsible for the damage. He set the camera to record at a time interval of between three to six seconds, figuring that this would be suitable to catch good footage of the perpetrators.

Over the coming years he captured a great deal of strange footage—some of it showing small beings that appeared to be coming into his back yard via some kind of a transporter beam. These beings, which he began to refer to as "floaters," resembled small spacemen wearing what looked to be spacesuits and helmets. In some photos they appear to be stacked one on top of the other inside the transporter beam, and in certain shots their tiny hands can be seen holding onto some type of pole.

Greg also managed to record many balls of light with different shapes and sizes, including one that appeared with a flat bottom. His footage also revealed several strange creatures that appeared to walk upright and have long tails, as well as having clear mouth and eye features. In one series of photographs, Greg is outside at night in his garden with his young daughter, when all of a sudden a large chair-like object appears directly over one of the shrubs in the garden next to them. The object then drops down towards the ground at the same time it is rotating. Although Greg and his daughter were looking in the direction that the object appeared, they saw nothing and were totally oblivious to its presence.

In another photograph the image of his garden appears to have been split vertically. As a result, half of Greg's garden seems to have disappeared, being replaced by another image that looks like it could be from a completely different time and place. In some instances, energy spikes similar to lightning have been captured in his footage, as well as several entities that appear to be disguising themselves in some way.

Greg continued recording for a number of years and gathered a massive collection of footage showing the strange goings on in his back yard. I found the whole story very interesting and have read his book *The Alien Connection* a number of times, and watched his DVD on many occasions. There is some great footage on the DVD showing the "floaters," as well as examples of other strange phenomena that Greg captured on his security camera—all of it offering more proof of a hidden reality.

All told, my research, combined with that of others, has left many questions unanswered. That is why I must carry on with my quest into the invisible. I expect it will take several more years of experimenting to arrive at some clear answers, especially since I am working solo. In this coming summer (2016) it is my intention to begin filming using the infrared-enabled Canon G10 in movie mode, combined with my new full spectrum HDR-PJ620 Sony Handycam and an ultraviolet pass filter. I will also use another Canon standard digital camera in movie mode, and set them all up

Conclusions

to film exactly the same scene in a series of proper, controlled experiments. This way I will be able to capture clear footage of the infrared, the ultraviolet, and the visible light spectrum all at the same time.

I also plan to have a radio playing in the background, which will make it easier to pinpoint the exact moment when an invisible UFO flies past, and maybe this will help provide more evidence to back up my photographic work. I intend to put the findings on my Quest for the Invisibles YouTube channel. From past experience, the skyfish and other forms I have captured in the ultraviolet and infrared are not present on simultaneous footage taken with a standard camera, filming the visible spectrum.

As I have recorded much closer and clearer footage in the ultraviolet, I will continue to record with the Handycam using the sun obliteration technique, with the infrared and standard cameras set to film the same scene from exactly the same angle. I will also try to determine the true size of some of the invisible life forms including skyfish, and intend to place a series of objects along the apex of the roof being used to block out the sun. This will help determine whether these invisible forms are going in front of or behind these objects when they pass over the top of the roof, and therefore give me more of an idea as to their dimensions.

I will be concentrating on the infrared using my infrared-enabled Canon G10 in movie mode, as well as using my full spectrum Handycam with an external infrared pass filter. I will record exactly the same scene, but will set up both the camera and the Handycam to record at different angles, and this way, different views of the same invisible objects may be provided. I also plan to record at night using my infrared CCTV camera in combination with the full spectrum Handycam with an external infrared pass filter, as well as the infrared-enabled Canon G10. This will reveal the same nighttime scene using three different recording devices at exactly the same time.

The advantage of having the infrared illumination beam is that both the Handycam and the Canon G10 will also benefit from the infrared light coming from the CCTV camera, enabling them both to take much clearer images. Any subsequent invisible forms that

Quest for the Invisibles

pass through the infrared illumination beam will appear much brighter, and this will really help with the clarity of the footage. I will position the three recording devices at slightly different angles to reveal a variety of views, and again, will have a radio playing in the background to help pinpoint the moment when any potential objects pass by. This will also help prove that all the footage was taken at exactly the same time. I will be pretty busy over the next few years with these experiments, but it is something I enjoy and look forward to doing.

I hope this book has given you an insight into the invisible world around us, and maybe one day you might want to take a glimpse into this strange borderland and discover for yourself just what is out there. If this happens to be the case, then I wish you good luck.

Appendix 1

Infrared Photographs of Invisible UFOs

The following photographs and video stills were taken between mid-April 2010 and November 2015 during daylight hours, using an infrared-converted Canon G10 with an internal 720nm infrared pass filter. All of the footage taken on the G10 was filmed at 30fps in and around my research area, which is situated just a few miles outside of Oxford, UK.

I began filming in early February 2010; which saw most of the UK still gripped by ice and snow. In mid-April 2010 I started to use a Reich cloudbuster as an attraction method and began to capture my first bits of infrared footage. It was at this time I began using the Canon G10 on the movie mode setting, aiming the camera exactly where the cloudbuster was pointing, as well as walking around my research area taking multiple photographs when I felt the urge to do so.

The objects that I managed to capture varied tremendously. Many appeared to be life forms that were moving with purpose through the atmosphere, such as various types of skyfish and a whole host of other strange invisible forms, including many craft-like objects. Nothing was ever visible at the time of taking the photographs, although many were objectified using techniques for "seeing" the invisible, as described in the book.

The last six photographs in this section were taken from footage filmed at twenty-five frames per second using my CCTV camera, as well as the full spectrum Handycam with an infrared pass filter and a UVR filter. The two pieces of footage that were taken using the CCTV camera were filmed at night.

There are a whole range of external infrared pass filters available in many different sizes, and these can be used with full spectrum cameras or camcorders. The filters block the visible light spectrum while passing the invisible infrared radiation at many different points within the near infrared spectrum.

 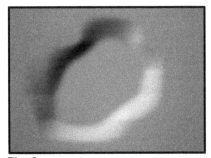

Fig. 1   **Fig. 2**

Fig. 1 shows the visible UFO that I encountered on Christmas day evening back in 2008. The UFO was moving from left to right, and a slight blur can be seen on the left side, which is a result of this movement. **Fig. 2** shows an embossed version of the same photograph, which has helped to bring out the overall shape and body curves just as I remember them as the object flew directly above me at little more than a hundred feet or so, before heading over the rooftops of a nearby housing estate.

I followed the UFO for quite a while until it came to a stop above some trees in a local park. At this point I decided to go back home and get my camera. When I finally got back to the park there was no sign of the UFO, so I returned home once more. Just as I was about to head back down to the park a third time, I caught another glimpse of the object in the distance as it emerged from the thick fog. I was then able to capture an image on a few seconds of movie mode footage, before it completely disappeared from sight.

This video still was taken on my standard Casio Exilim digital camera using the movie mode setting, and I still count myself very lucky to have captured it due to all the branches and obstacles in the way that evening. This would ultimately lead me on a quest to obtain more footage, and I eventually discovered the pioneering work done by Trevor James Constable in the early 1950s, when he managed to photograph invisible UFOs in the infrared part of the spectrum.

As a result of reading the books written by Constable, I obtained a digital camera, which I then sent away to a company in the USA for an infrared conversion. Soon after, I began to capture my own footage of invisible UFOs.

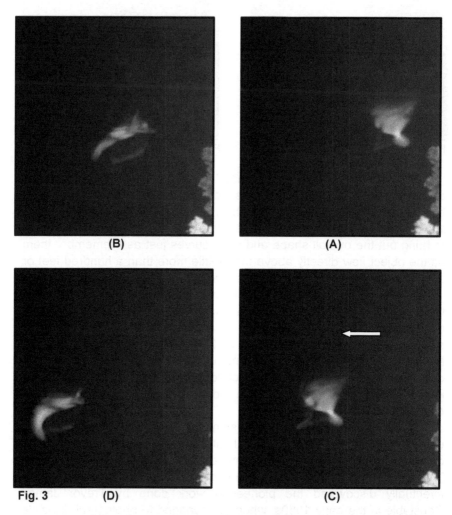

Fig. 3 A–D shows four consecutive frames taken of a "spinning jenny" as it moves through the atmosphere (note direction from arrow). Footage is from the Hidden Reality YouTube video "Spinning Jennys & Plasmatic Gliders," and was recorded at 30 fps on October 2nd 2011 at 4:05 pm, using my infrared Canon G10 in movie mode. The fish-like body and tail can be seen between each "spin" and the right wing can just be made out, although it is hardly reflecting any infrared radiation in this position. When these life forms are filmed in full spectrum, only a ghostly see-through image of light brown whirling energy can be seen in the footage.

Fig. 4

I have always referred to this photograph as the "Feather Critter" because of its similiararity to a feather. This example of invisible life was captured on April 2nd 2011 at 12:19 pm at $1/60^{th}$ sec. I was just walking past the tree and suddenly felt the urge to take a few photographs after I thought I could "see" something using techniques described in the book. The object appeared in only one photo of the three that were taken of the area just above the top of the tree, which is situated in one of my neighbours gardens.

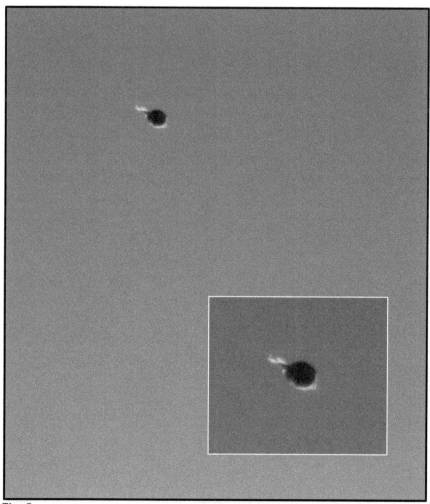

Fig. 5

This is one of my favourite infrared photographs, and it was taken just above a large bush in my garden. This is the only example I have of this type of "invisible," and I consider myself very fortunate to have caught it on the infrared G10. The photograph was taken on March 4th 2011 at 3:10 pm on a very bright afternoon at 1/2050th sec., and was the second of three photographs taken in quick succession. Nothing appeared at all on the first and third photographs taken of exactly the same scene.

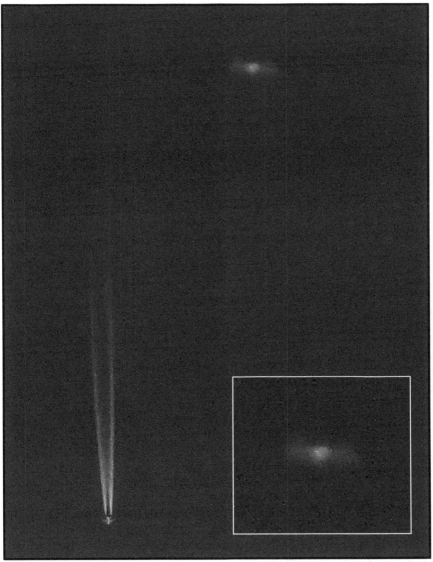

Fig. 6

This invisible UFO was captured on June 26th 2011 at 11:49 am. I took multiple photographs of the plane as it flew overhead, and this object appeared in the third photograph, taken at 1/400th sec. There was no sign of it in any of the other photographs taken that morning.

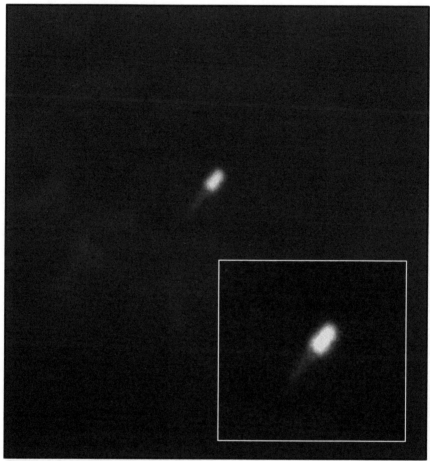

Fig. 7

A plasmatic-looking craft in obvious motion shoots skywards at colossal speed, leaving a small trail which can be made out at the rear. This photograph was taken on a dark, stormy day on June 27th 2010 at 12:36 pm. It was one of a handful of shots taken of the general area where the cloudbuster was aiming at the time.

The object only appeared on the second of the three shots taken at 1/10,000[th] sec., and there was no sign of it on any of the other photos taken.

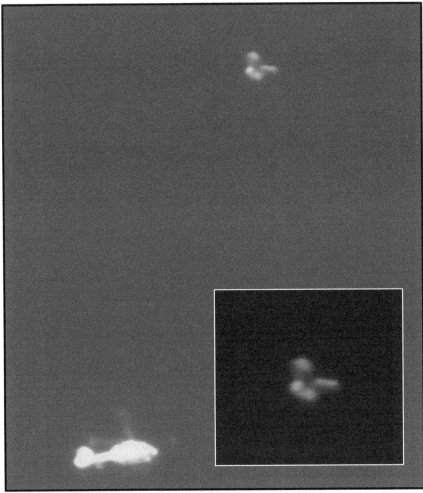

Fig. 8

Here we see a helicopter passing over my research area, reflecting an incredible amount of infrared radiation from the warm November sun. I took several photographs as the helicopter passed overhead, and this strange looking "invisible" appeared on the first shot (of the many taken) at 1/60th sec, on November 1st 2015 at 2:05 pm.

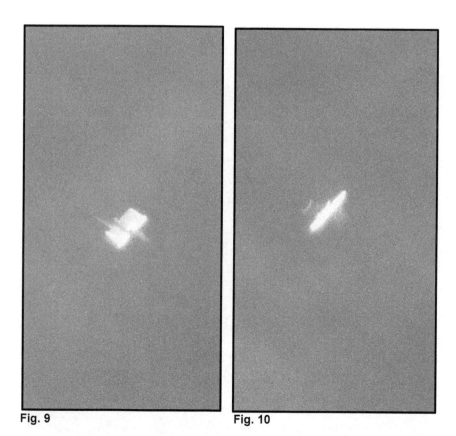

Fig. 9 Fig. 10

    I photographed this invisible craft-like object right above my research area on 8th September 2010 at 6:20 pm after suddenly getting the urge to take a few more photographs of the area where the cloudbuster had been pointing before packing up for the day. I lifted the infrared camera over my head and took four shots, one after another, at 1/15th sec.

    Upon viewing the first photograph I assumed that the white objects on either side of the main body were wings and that it was moving forward, much like an aeroplane. In the second photograph only a small section of the wing-like structure was visible, and the main body had all but disappeared from view. It was in fact travelling from about 8 o'clock to 2 o'clock, lower left to upper right, and not in the sideways direction I had assumed. The third photograph taken looked identical to the second, but by the fourth photograph, the UFO had completely disappeared from view.

Fig. 11

Fig. 11 shows an invisible UFO moving just above the roof of one of my neighbours houses, just after noon on March 21st 2014. This video still was taken using the Sony Handycam, fitted with a 780nm infrared pass filter. This footage was taken shortly after I first purchased the filter while still experimenting with the sun obliteration technique.

Fig. 12

This is another video still taken using the Sony Handycam, fitted with a 780nm infrared pass filter. This strange looking invisible object was also captured using the sun obliteration technique, just as it was passing above the roof of my garage on March 21st 2014 at 12:30 pm.

Fig. 13

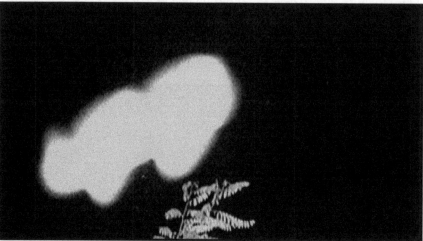
Fig. 14

The above video stills were both taken from nighttime infrared CCTV footage which was filmed from a small room situated above my garage. In **Fig. 13** we see an almost amoeba-like invisible life form, captured as it shoots down towards the ground at an immense speed during a freezing cold evening on February 23rd 2013 at 9:25 pm.

Fig. 14 shows a rather strangely shaped invisible form, captured as it moves from left to right just above the treetops and towards the rear of my garden. This footage was taken on September 12th 2015 at 9:52 pm.

Fig. 15

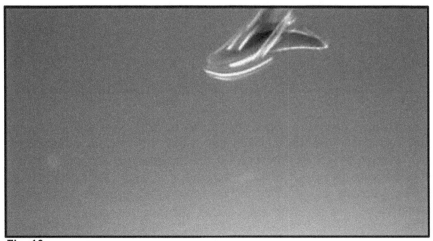
Fig. 16

In **Figs. 15** and **16** we see a rather peculiar looking invisible sky creature, captured just above the roof of my house on March 16th 2014 at 2 pm, using the Sony Handycam fitted with an X-Nite UVR filter. This filter blocks the visible light spectrum, while passing its two invisible ends i.e. the infrared and the ultraviolet.

In **Fig. 15** the life form appears to be flapping frantically, and in the next still we get to see the actual body shape showing the tail and small wing-like appendages which have now become visible.

Fig. 16

Nik Hayes sets up the cloudbuster at his research area near Oxford, in early July, 2012. The Reich cloudbuster is built from the basic inventions of Dr. Wilhelm Reich. Its operation excites the ether and appears to affect other densities of existence. Aside from being able to dissipate clouds, it is also capable of attracting UFOs—both craft and their biological counterparts.

The cloudbuster is basically a row of pipes that can be aimed in any direction and elevated at any angle. Any number of pipes can be used, and they are ultimately grounded in water at one end. Trevor James Constable built many different cloudbusters, which he initially used for attraction purposes while filming invisible UFOs, but later began using them for his experiments in weather engineering.

After a relatively short period of time the pipes of the Hayes cloudbuster had become so magnetised that they could easily spin the needle of a compass right around in either direction, severely affecting its ability to detect true north and give accurate readings.

Appendix 2

Ultraviolet Camcorder Stills of Invisible UFOs

The following footage was taken during the daytime, using a full spectrum Sony DVD650 Handycam with an external X-Nite 330nm ultraviolet pass filter. The Handycam, which films at twenty-five frames per second, was attached to a tripod, while using the sun obliteration technique to create what I refer to as an "illumination zone."

Once anything entered this zone it would be completely illuminated by the sun's rays and thus become visible to the lens of the camcorder. Very often the objects would fade out and become barely visible once they pass this zone, but I would sometimes encounter objects that were self-illumed and didn't require the sunlight to illuminate them.

Fig. 17

My first capture in the ultraviolet spectrum was similar to the example that we see above, in **Fig. 17**. It was moving what looked like tentacles and appears to bear a striking similarity to a jellyfish in its general body shape. The footage above was taken on September 29th 2011 at 2:46 pm, as the object was travelling just above the roof of my house. No attraction method is needed in order to capture this type of footage.

Fig. 18   **Fig. 19**

In the above video stills we see a perfect example of an invisible object entering the illumination zone and forming. In **Fig. 18**, the mouth-shaped part has already become visible to the ultraviolet lens, and in **Fig. 19**, taken only a fraction of a second later, the rest of the mushroom shaped body is revealed. This footage was taken on September 29th 2011 at 3 pm, just above the roof of my garage.

The sunlight can be seen lighting up the edge of the roof and creating the illumination zone. Anything moving into or through this point will be completely illuminated by the sun's rays, and therefore will become visible to the ultraviolet lens of the Handycam for a fraction of a second.

Once past this point, most objects fade out and can hardly be seen at all, eventually becoming invisible once more.

Fig. 20

This video still shows one of my stranger captures in the ultraviolet and was taken on an extremely hot day on July 31st 2013 at 12:10 pm. The object was travelling from right to left and appeared in two frames, although by the second frame only the topmost tip of the propeller-like feature was visible in the photograph.

Fig. 21

Fig. 21 was recorded on July 21st at 12:19 pm, just as the object was passing above the roof of one of the gable ends of my house. It is totally illuminated by the midday sun, which can be seen just rising above the edge of the roof. The atmosphere is full of invisible life forms such as this, and anyone can capture similar footage using a full spectrum camcorder and a UV pass filter.

Fig. 22

Shown is a peculiar looking fish-like form consisting mainly of a mouth which extends for most of its body, and a small tail-like feature. This specimen was captured as it travelled from right to left at an incredible rate near to the apex of the house roof, on Sept. 7th 2012 at 6:45 pm.

As with quite a few of the invisible UFOs that I have captured in the ultraviolet part of the spectrum, many appear to be riding on the west to east energy flow which I first became aware of during 2011, with some of them travelling backwards as they shoot through the atmosphere.

This example, which appears to lack any fins or wing-like structures, seems to be using the tail-like feature to help steer itself as it travels through the atmosphere, and can be seen curling around to the right of the main body. The small dots of light that can be seen behind the object are tiny insects that are also reflecting ultraviolet light.

Many of these objects are travelling at extreme speed and only appear on one or two frames, so it is important to slow the footage way down when reviewing it to cut down the likelihood of missing anything. On many occasions I have gone over old footage only to find something that was overlooked when I had originally checked it. It is advisable to always take your time and go over the footage a few times to be sure.

Fig. 23

Fig. 24

This example was filmed above the roof of my neighbour's house on May 23rd 2013 at 7:26 pm. In **Fig. 23** the object is directly in the illumination zone and is completely lit up. In the next frame, **Fig. 24**, the object has already started to fade as it leaves the illumination zone and the flap at the rear has changed position. The tongue-like protrusion at the front of the object, which is evident in both stills, seems to have grown much darker but remains in about the same position as in **Fig. 23**.

This is one of the strangest objects I have come across since I first began exploring the ultraviolet part of the spectrum, and was totally lost for words when I first saw the images after loading the footage onto my computer. The variety of invisible forms out there never ceases to amaze me, and just when you think you have seen it all, another even stranger looking object comes along and makes its presence known.

Fig. 25　　　　　　　　　　　Fig. 26

In **Figs. 25** through **27**, we see three consecutive frames showing another invisible form as it moves left to right above some nearby garage roofs on April 20th, 2013 at 3:40 pm. The main body of the sun is hidden just below the roofline, and in **Fig. 27** it can be seen lighting up some cobwebs on the edge of the roof.

Instead of using the normal "sports" setting on the Handycam I decided to use the white balance selection which I had already set prior to filming. This, however, made the sky look slightly violet in colour, giving many of the invisible objects a whitish, almost see-through complexion. I later reverted back to using the automatic selection which I normally use, as I felt this gave the footage a bit more clarity and tended to bring out more detail.

I have captured similar objects to this one in earlier footage, but this proved to be the best example as it proceeded to turn around mid-flight, allowing us to view it from three different angles and revealing the tube-like structures evident in **Fig. 26**.

Fig. 27

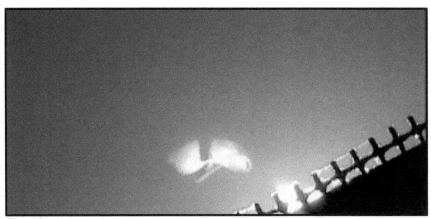

Fig. 28

In **Fig. 28** we see another object passing through the illumination zone, this time just above the guttering of the summer house, towards the rear of my garden. This video still was taken on July 19th at 7:14 pm, which had been a particularly hot day—and by the next frame, only a very small part of the wing-like protrusion could be seen in the footage.

Fig. 29

Fig. 30

This footage was taken on May 26th 2014 at 12:40 pm, and shows a most bizarre looking invisible UFO moving just above my house roof. In **Fig. 29** the object is just entering the illumination zone, and certain features are becoming apparent. In the next frame it is now totally illuminated, and the rod-like protrusions have become more visible.

Fig. 31

Fig. 32

Fig. 33

In **Figs. 31** through **33**, we see three more stills taken using the full spectrum Handycam with an external 330nm UV pass filter, showing an invisible flying life form heading towards the gutter protection on the edge of my house roof. There are features that can be made out on the first two stills such as the wing-like shapes, and an almost corrugated looking body part. This footage was taken on another extremely hot summer day back on July 19th 2013 at 1:15 pm.

By **Fig. 33** the shape has completely changed and almost resembles a guitar in appearance, and the only thing visible in all three shots is the handle-like protrusion which appears more clearly on the last photograph. Either this life form is morphing as it moves, or certain parts are not reflecting the ultraviolet radiation as well, as it leaves the illumination zone.

I don't normally allow so much sunlight into the scene as it can cause a "white out," and I was just about to adjust the position of the Handycam when I saw this object shooting down from the sky on the LCD viewing screen.

Fig. 34

Fig. 35

Fig. 36

In **Figs. 34** through **36** we see another example of a shape-shifting invisible UFO captured as it shoots by, diagonally, from right to left, just above the main roof of my house on June 4th 2013 at 4:12 pm. The sun can be seen behind the gutter protection and has completely illuminated the object in all three video stills.

There is a fine line between zooming in too far with the camcorder, and not zooming in enough. If you zoom in too much you may catch the object in only one frame, and if you do not zoom in enough you may catch the object in multiple frames, but the size of the subsequent images may be far too small to see any significant detail.

This particular footage was taken one afternoon while I was trying out a new tripod that I had purchased only a few hours before. I was experimenting with different angles of filming at the time, and had only been outside for around five minutes when this strange looking form showed up on the footage.

I have filmed similar objects to this one before, but have captured them in only one frame. This object appeared in three frames, giving us the chance to see the object's progression as it moved through the atmosphere.

Appendix 3

Ultraviolet Camcorder Stills of Skyfish

The following video stills were taken during 2012 and 2013 using the Sony DVD-650 Handycam, which was fitted with an external X-Nite 330nm ultraviolet pass filter. They were recorded using different parts of my house roof to block out the main body of the sun—basically, the sun obliteration technique. The images here represent a small amount of the total footage recorded during this time period but, in my opinion, are some of the best examples I've captured of these fast moving, invisible life forms.

On most occasions I recorded exactly the same scene using my standard Casio Exilim digital camera which was set to movie mode, but no images of skyfish ever appeared in the resulting footage. If you look carefully, some of the images contain small specks of light, and these are, in fact, insects that were captured in the footage along with the skyfish. As you can see, they are also reflecting ultraviolet light, but appear quite small in relation to the skyfish. What I have noticed during my time spent recording in the ultraviolet part of the spectrum, is that most insects appear like this and look nothing like the skyfish. The camcorder is focused on the edge of the roof and zoomed in towards it, and the majority of insects are barely noticeable due to their limited body mass.

In most cases, the only parts of the skyfish that are reflecting the ultraviolet light are the fins or tail-like features, whilst in most cases the middle body section remains barely visible, even to the UV lens of the camcorder. Most of the skyfish I've captured in the ultraviolet seem to be moving on the west to east etheric flow, which goes from right to left above the main rooftop of my house. This otherwise invisible flow is why most of my footage shows them moving in the same direction. Most insects on the other hand move up and down, in many different directions, and seem totally unaffected by this energy flow. On many occasions I have seen them fly right through it as if they are totally unaware of its existence.

Fig. 37

Fig. 38

Fig. 39

Fig. 40

Fig. 41

Fig. 42

Video Stills of Insects Elongated by the Filming Process

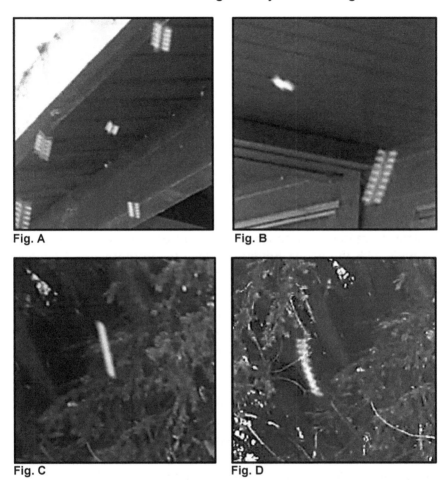

Fig. A

Fig. B

Fig. C

Fig. D

Fig. A shows a group of mosquitoes. **Fig. B** shows a midge. **Fig. C** shows a small moth, and in **Fig. D** we see a medium sized crane fly. This footage was taken during the afternoon on March 12th and April 13th 2016, and shows insect species that are typical inhabitants of an English country garden. These images were filmed using a hand held Casio Exilim digital camera in movie mode. The insects were around a foot or so away from the camera lens when the footage was taken, although on a couple of the recordings I had also zoomed in a little. The multiple wing effect caused by elongation can clearly be seen in Figs A, B, and D.

As seen from the video stills found on the previous page, images of insects can be elongated by the photographic process. These are a product of the video recording, but can sometimes be caused by a long exposure time when taking a photograph at night or in a low light situation. Similar footage and photos of insects have been presented as rod/skyfish in the past by others, and it is easy to see how this mistake can be made. At first glance, the insects presented do give the impression of a rod-like creature, and it is not always easy to tell that they are, in fact, photographically-elongated insects. Such misinterpretations have not helped in the rod/skyfish debate, but it does not mean that all of the footage presented can be dismissed as stretched out insects. Many examples of skyfish do not look anything like these.

When the same insects are filmed out of direct sunlight, the repeated wing effect diminishes sharply due to the lack of reflected sunlight. In many such cases all that can be seen is a dark, rod-like body with the wings hardly visible. I found that when I filmed the insects from twice the distance away or more, there was no noticeable elongation and most looked nothing like the examples shown. In some cases, all that appeared in the footage was a white rod shape—as found in Fig. C, which shows a small moth as it goes about its business. The multiple wing effect was less visible as the insects moved further away from the camera lens, until eventually it diminished completely.

I have seen some footage by Jose Escamilla from the trailer of his new film, and in this he describes how he has now captured skyfish using a high-speed, non-interlaced video camera. There are several shots showing various images, some of which I assume were filmed using the high-speed camera. If this is so, then none of the images should have been stretched in any way, due to the very high speed. Under closer examination the footage revealed itself to be very similar to other footage I have seen by Jose and it is also similar to footage I have captured in the infrared over the last few years.

Many people are too quick to jump on the bandwagon and dismiss things at the first opportunity, without giving others the chance to fight their case properly. As soon as it was found that insects could be stretched by the recording process, many took this revelation as being absolute proof that rods/skyfish do not exist, and they refused to believe otherwise. The entire subject was then treated as if the case was closed, and that was the end of that. I have seen the same reaction to the subject of UFOs and crop circles, both of which have been featured prominently in the public eye over the years. It seems there are people in high places who are desperately trying to keep the truth away from the general public.

Jose Escamilla himself has been the target of campaigns mounted to discredit him and his findings, but he keeps exploring the subject and collecting more and more evidence. In my opinion he is a pioneer and has devoted his life to proving that skyfish are a real phenomenon. I only discovered them accidently as I filmed for invisible UFOs from my research area, but over the last five or six years I have seen more than enough footage to convince me that they are real and do exist.

Most winged insects in flight will probably look similar to the stills I have provided, especially when they are flapping their wings quickly, trying to gain height or momentum and are in close proximity to the camera. However, when much further away or higher in the sky you won't get the multiple-wing effect caused by elongation. They will look like normal insects, although much smaller in size due to the distance away. Skyfish don't do this.

A larger flying beetle or a bee will look slightly different when it is elongated by the filming process, as would a dragonfly, but any piece of footage taken of an insect with its wings flapping and reflecting sunlight will give a similar image to the ones included here, I imagine; although It could be possible that the wings might project out at a slightly different angle, giving a more v-shaped appearance than is shown on the examples. The image would also look slightly different if filmed from a more sideways view, or from an angle where the insects' wings weren't reflecting so much sunlight. In cases with insects being slightly further away from the lens and reflecting lots of sunlight, the multiple-wing repeats would sometimes appear as an almost pure white block, where no individual wing was visible.

Over the years I have seen lots of skyfish footage that is quite different from anything that could be formed from an image of an elongated insect. There are certain features found on some skyfish that do not appear on any of the insect shots I have ever taken or ever seen on other people's footage. The larger skyfish, including the missile/torpedo types, appear to have at least two pairs of long, paddle-like appendages located at either end of their bodies, and normally show one on either side, towards the centre of the body. I have sometimes seen two smaller pairs situated at the front and back as well, appearing just in front of the larger ones. I can't imagine how insect's wings could ever look like these long, paddle-like appendages, especially as the larger ones stick out quite far from the main body and look totally disproportionate when compared with the size of an insect's wings in comparison to its body.

The domed front ends that appear on some large skyfish can make them look very much like a missile or a torpedo, especially when combined with the long, thin body shape. These are features that are not apparent on any insects I have ever seen or filmed. Some examples of skyfish I have recorded in the infrared have been quite short and had sharp pointed ends, while others have been very long and look much like javelins. In some of my footage they can be seen flying out from behind trees towards the back of my research area, a distance of at least twenty feet away. This reference point is a great help in determining their distance from the camera. Any insect filmed from that kind of distance would hardly be visible to the camera, and it certainly would not be elongated by the recording process, as it is much too far away from the lens.

I tried, on many occasions, to recreate the elongated insect footage shown on page 145 using my infrared Canon G10. What I found was that the insects didn't have the same clarity when filmed against the backdrop of white foliage—created by the reflected infrared radiation. At times, they could not be seen at all despite their wings being illuminated by the sunlight. Even the wood of the garden shed changed colour and became much lighter when viewed through the lens of the G10, as did many other things such as brick walls and even the sky. Using the standard Casio Exilim camera, it was found that the insects were much more visible and clear against a normal coloured backdrop of green foliage and a blue sky, and the elongated effect could be seen with much more clarity.

There are many reasons to believe that skyfish are much more than just a product of the way insects can be recorded, and it is my aim to continue gathering evidence to show this to be the case. I will continue to film as many other invisible life forms and objects as I can in both the infrared and the ultraviolet. It is only through research and constant study that the answers we seek will be found.

Bibliography

The Cosmic Pulse of Life – Trevor James Constable – 2008
Published by The Book Tree
ISBN 978-1-58509-115-7

They Live in the Sky – Trevor James Constable – 1958
Published by Saucarian Books (out of print)
Second-hand copies are still available over the Internet

UFO: La Realta Nascosta – Luciano Boccone – 1980
Published by Ivaldi Editore (out of print)
Second-hand copies are still available over the Internet

The Flying Saucers are Real – Donald Keyhoe – 2006
Published by The Book Tree
ISBN 978-1-58509-264-2

Gods of Aquarius – Brad Steiger – 1982
Published by Berkley (out of print)
ISBN 978-0425055120
Second-hand copies are still available over the Internet

The Alien Connection – Gregory Harold – 2008
Published by Dog Ear Publishing
ISBN 978-159858-856-9
Available from Amazon.com
Can also be ordered from any good book store

Harold's Mystery – DVD – Gregory Harold
Available from Amazon.com
Can also be ordered from any good book store

ALSO AVAILABLE FROM THE BOOK TREE

THE COSMIC PULSE OF LIFE, by Trevor James Constable. Presents evidence that UFOs are mainly invisible and consist of both physical craft and living biological creatures. The author convincingly shows that our atmosphere is the home of huge, invisible living organisms that are sometimes confused with spacecraft when they become visible. Mr. Constable has photographed both types of UFOs with infrared film, some of which are reproduced in this expanded, updated edition. Despite his bold leap into the future, the general public and official ufology have a hard time accepting the evidence. In more recent years, teams of engineers and technicians in both Italy and Romania, unaware of Constable's earlier discoveries, obtained virtually identical infrared photos, published in Italy. In 1996, NASA used ultra-violet sensitive video to record swarms of invisible UFOs that looked like Constable's earlier photos. Examples are in this book. Also covered are earlier pioneers in important life energies that play a big role in this research, including Dr. Wilhelm Reich, Rudolf Steiner, and Dr. Ruth B. Drown. Recommended for those interested in UFOs and higher realms of physical reality. **364 pages, 6x9 paper, $29.95**

HIDDEN HISTORY, RAIN ENGINEERING AND UFO REALITY, by Trevor James Constable. This tribute book covers three subject areas that span over a half-century in the life of Trevor James Constable, who passed away in April, 2016. The hidden history area involves little-known military men and their amazing heroics. The rain engineering section documents the evolution of the technology along with the resulting proof of success the author experienced from around the world. If taken further, this technology could be a tremendous help to humanity. Lastly, the reality of UFOs covers flying, biological creatures that were found to reside in the hidden ultraviolet spectrum of our skies. When witnessed, these beings can be mistaken for unknown craft that are thought to be piloted by humans or aliens rather than being alive in their own right and resident to the earth. The pioneering work of Constable is still ahead of its time. The compelling scientific proof presented here leaves little doubt that we have much more to learn about ourselves and the world around us. **260 pages, 6x9 paper, 22.95**

THEY LIVE IN THE SKY, by Trevor James Constable. This book probes into the question of extraterrestrial life and craft from an entirely different angle, experienced first-hand by the author. Most known for his ground-breaking infrared photography of objects in our skies, this earlier work describes his own telepathic contact with an invisible being. It was from the suggestions made by this entity that Constable began his work with infrared photography. The results shocked him so much, and he was so unsure as to how such events would be perceived, that he wrote the book under a pseudonym, Trevor James. In it, he makes no claims about having any physical contact with beings or craft, but effectively advances the theory of there being invisible craft and creatures all around us, residing in our own atmosphere. New methods of investigation were first proposed in this book, and it should be required reading for any serious researcher who is unafraid to explore new areas. **312 pages, 6x9 paper, 24.95**

TO ORDER PLEASE VISIT www.thebooktree.com OR CALL 1-800-700-TREE (8733) 24 hrs.

Subscribe to FATE Magazine Today!

- Ancient Mysteries
- Alien Abductions
- UFOs
- Atlantis
- Alternative Archaeology
- Lost Civilizations
- And more ...

FATE covers it all like no else. Published since 1948, FATE is the longest-running publication of its kind in the world, supplying its loyal readers with a broad array of true accounts of the strange and unknown for more than 63 years. FATE is a full-color, 120-page, bimonthly magazine that brings you exciting, in-depth coverage of the world's mysterious and unexplained phenomena.

1-year subscription only $27.95
Call 1-800-728-2730 or visit *www.fatemag.com*

Green Subscriptions. E-issues.
Only $39.95 for the entire year!
Go green. Save a tree, and save money.

12 issues of FATE delivered electronically to your computer for less than $3.95 an issue. • Receive twice as many issues as a print subscription. Includes six regular issues plus six theme issues (UFOs; Ghosts; Cryptozoology & Monsters; Nature Spirits & Spirituality; Strange Places & Sacred Sites; and Life After Death). • Free membership in FATE E-club (save $10). • Free all-access to Hilly Rose shows (save $12.95). • Members-only video interviews. • Discounts on all FATE merchandise. • Monthly Special Offers.

Lightning Source UK Ltd.
Milton Keynes UK
UKHW02f0614060618
323809UK00010B/479/P